Ocean Circulation and Pollution Control –
A Mathematical and Numerical Investigation

Springer-Verlag Berlin Heidelberg GmbH

Jesús Ildefonso Díaz (Editor)

Ocean Circulation and Pollution Control – A Mathematical and Numerical Investigation

A Diderot Mathematical Forum

Springer

Editor:
Jesús Ildefonso Díaz
Departamento de Matemática Aplicada
Facultad de Matemáticas
Universidad Complutense de Madrid
28040 Madrid
Spain
e-mail: ildefonso_diaz@mat.ucm.es

Cataloging-in-Publication Data applied for
A catalog record for this book is available from the Library of Congress.

Bibliographic information published by Die Deutsche Bibliothek
Die Deutsche Bibliothek lists this publication in the Deutsche Nationalbibliografie;
detailed bibliographic data is available in the Internet at <http://dnb.ddb.de>.

ISBN 978-3-540-40647-1   ISBN 978-3-642-18780-3 (eBook)
DOI 10.1007/978-3-642-18780-3

Mathematics Subject Classification (2000): 35QXX, 86A05, 65-XX, 91-XX

http://www.springer.de

© Springer-Verlag Berlin Heidelberg 2004
Originally published by Springer-Verlag Berlin Heidelberg New York in 2004

Cover design: *design & production,* Heidelberg
Typesetting by the authors using LaTeX
Printed on acid-free paper   41/3142ck-5 4 3 2 1

Preface

In the framework of the Diderot Mathematical Forum (DMF) of the European Mathematical Society (EMS), a videoconference linking three teams of specialists was held in Amsterdam, Madrid and Venice on December 19th and 20th, 1997. The general subject of this videoconference —which was the second one of the DMF series— was *Mathematics and Environment: Problems related to Water*. The large number of common problems that were treated in the three sites highlights the global nature of environmental studies.

This book contains the written contributions assembled in Madrid. It is not surprising that a large part of this material focuses on questions related with Oceanography, since Spain is one of the European countries with longest coast.

In the contribution by R. Bermejo, ocean circulation is considered from the point of view of a new algorithm for the numerical approach to the so-called ocean primitive equations. A different perspective is followed in the paper by J. Macías, C. Parés and M. J. Castro. They consider a "local problem" whose great relevance goes beyond Spain: the Strait of Gibraltar. Important concrete aspects must be correctly modelled by means of a model obtained under some simplifications: the multilayer shallow seas model. The numerical experiences presented in this article allow to understand certain phenomena that have been known empirically since long time ago. A related question was considered by B. Sommeijer in Amsterdam. The use of sophisticated numerical algorithms was also the aim of one of the contributions presented in Venice (by A. Quarteroni).

One of the main environmental problems related with water concerns contamination and its many different aspects. Groundwater pollution is considered in the paper by M. W. Saaltink and J. Carrera (and was also discussed in the lecture by F. J. Elorza). This type of problem was, in fact, the main subject developed in the presentations made in Amsterdam by C. J. van Duijn, M. de Gee, B. H. Gilding and A. Stein.

The contamination of waters has many common points with the dispersion of air pollulants from combustion of fuel, as presented in the article by G. Winter, J. Betancor and G. Montero (and also by L. de Haan in the set of Amsterdam lectures).

The modelling of the interaction between physical and biological aspects in coastal ecosystems has a crucial relevance near cities and beaches. Such problems are considered in the paper by A. Bermúdez, C. Rodríguez, M. E. Vázquez-Méndez and A. Martínez. Since this problem is especially important in places such as Venice, it is understandable that the subject was extensively taken into account by Italian lecturers (A. Bergamasco, V. Casulli and G. Gambolati). It was also considered by B. Sommeijer in Amsterdam.

The above-mentioned problems are modelled in terms of systems of non-linear partial differential equations. Nevertheless, the study of environment very often requires methods that come from statistics. The use (and abuse) of statistics in environmental issues was the main content of several lectures given in Amsterdam (by L. de Haan, R. D. Gill and A. Stein). Finally, economical aspects and policies related with pollution limitations were taken into consideration by J. I. Díaz and J. L. Lions by studying the optimisation and control of multi-criteria under a deterministic state law. A different optimal control problem is studied in the paper by A. Bermúdez, C. Rodríguez, M. E. Vázquez-Méndez and A. Martínez.

Let us mention that several results contained in this book have not been published anywhere else. It is the case, for instance, of the contribution signed by the late Jacques-Louis Lions. Thanks to his dedication to this field in the last ten years of his life, the study of environmental problems reached a great popularity among mathematicians of many countries.

The videoconference was a complete success. The linking between the three sites worked perfectly and according to the programme. In fact, a video of more than 10 hours was recorded. As everybody knows, the organisation of an international meeting is always complicated. But this one was singularly difficult, due to the technical aspects required for a correct linking. The job of the three main organisers (E. Canestrelli, M. Keane and myself) was possible thanks to the help of a large number of people and institutions. Following a chronological order, we start by acknowledging the efficient work made by the EMS Committee on Special Events. Thanks to its President, Jean-Pierre Bourguignon —who was also President of the EMS at that time— and the Secretary, Mireille Chaleyat-Maurel, a joint preparatory meeting with the local organisers, held in Paris in May 1997, made easier the coordination and scientific design of the videoconference.

Green light was given by funds received from the DGXII of the European Commission. Support also came from other sources, such as the Unione Matematica Italiana and local institutions, especially the three universities linked by the event. We convey our indebtedness to all of them.

After the scientific design of the videoconference, the preparations required the help of many technicians and colleagues at each site. It is impossible to list all of them, but a special mention must be made to the high efficiency of Ben Schouten in Amsterdam, who helped not only with technical questions but also with many other aspects, including the scientific ones.

I cannot finish this short introduction without mentioning the important help received from Juan Francisco Padial, Carles Casacuberta (EMS Publications Officer), and the editors and staff of Springer-Verlag for the preparation of this book.

Jesús Ildefonso Díaz

The DMF Series

In 1996, the European Mathematical Society launched an original series of meetings named "Diderot Mathematical Forums". Here is how this choice came about:

- Since the ambition was to create a series of events during which mathematicians of different origins and specialists coming from other fields could exchange knowledge and discuss their views around a definite topic, why not call them forums?

- In the *période des Lumières*, Denis Diderot was one of the driving forces behind the writing and editing of *L'Encyclopédie*, an unbelievably ambitious entreprise exhibiting a new conception of the links between science, technique and society. He was also a man who always insisted in his writings on the human dimensions of things, and, in doing so, he was, in his own way, warning against "scientism", that was yet formally to be born. On top of that, by visiting different countries and working in them, he experienced Europe at a time where it was not so common; this makes of course another point for the relevance of his name in this context.

The main purpose of this series was (and still is) twofold. First, to provide mathematicians with a tool to change the image that their community projects outside in the wider society, by showing their interest in engaging themselves in exchanges with other professional groups. Second, to have an internal effect by giving more visibility to new avenues of thought for mathematicians, in particular young ones, with the definite purpose of showing how the interaction with other communities brings in new problems and sheds a new light on the mathematical practice and on the needs of other people.

This led the EMS to try out a rather unusual format for these Forums, namely having two-day meetings held in three different European cities linked through videotransmissions for a part of the event. The purpose at each location was to have manageable size audiences, say up to 100 participants, making the local organisation light enough while creating conditions for the thought for confrontation.

Very early, the theme "Mathematics and Environment" was identified as a topic to which a Diderot Mathematical Forum should be devoted. Indeed, this domain is exemplary of interdisciplinary actions that one will need to develop in the near future on a much larger scale if one is serious about coming to grips with complex systems such as this one. But it soon appeared that the domain was too wide to allow fruitful exchanges. This led the EMS committee in charge of the Forums to propose and concentrate attention on a narrower subject, namely environmental problems related to water. In this context, Amsterdam and Venice appeared as two emblematic European cities to discuss such questions.

The theme "water" speaks to every citizen, and it is especially appropriate for confronting the views of mathematicians with those of many other scientists (chemists, physicists, earth scientists, biologists, etc.). It touches upon serious societal issues, in the solution of which scientists of all sorts have yet to find their true place. This dimension, namely making the relevance of present day mathematics plain to a wider public, is part of the challenge that the Diderot Mathematical Forums want to face.

It is by now well recognised that there is still a lot to be done in the development of mathematical models allowing a better knowledge and prediction of water presence in the soil, together with pollution risks. In many European countries, and also in other continents, improving the production of usable water is very high in the priority list of problems to be solved. Many other issues related to water were addressed in the Forum, in particular those connected to sea behaviour. This topic was dominantly dealt with in the third city participating in the Forum, namely Madrid.

It is my pleasure to end this foreword by acknowledging the hard work done by Professor Ildefonso Díaz from the Universidad Complutense in Madrid, Professor Michael Keane from the Centrum voor Wiskunde en Informatica in Amsterdam, and Professor Elio Canestrelli from the Università di Venezia, to set up this Diderot Mathematical Forum. We are grateful to them for their hard work.

Jean-Pierre Bourguignon
EMS President 1995–1998
President of the EMS Committee on Special Events

Second Diderot Mathematical Forum

Mathematics and Environment: Problems Related to Water

Amsterdam, Madrid, Venice, December 19–20, 1997

Amsterdam

Organisers: MICHAEL KEANE AND BEN SCHOUTEN

R. M. COOKE (Technische Universiteit Delft)
Expert judgement and the theory of dry water

C. J. VAN DUIJN (Centrum voor Wiskunde en Informatica, Amsterdam)
Salt water intrusion in coastal regions

M. DE GEE (Landbouw Universiteit Wageningen)
Semi-numerical methods for groundwater contaminant transport

B. H. GILDING (Universiteit van Twente)
On the wetting front —transport of moisture in soil

R. D. GILL (Rijksuniversiteit Utrecht)
Lies, damned lies, or statistics of the environment; use and abuse of statistics in environmental issues

L. DE HAAN (Erasmus Universiteit Rotterdam)
Sea and wind: multivariate extremes at work

B. SOMMEIJER (Centrum voor Wiskunde en Informatica, Amsterdam)
Numerical modelling of three-dimensional bio-chemical transport in shallow seas

A. STEIN (Landbouw Universiteit Wageningen)
Point processes, random sets and geostatistics for analyzing patterns of methylene blue coloured soil

Madrid

Organiser: JESÚS ILDEFONSO DÍAZ

R. BERMEJO (Universidad Complutense de Madrid)
Eulerian versus semi-Lagrangian schemes in some ocean circulation problems: a preliminary study

A. Bermúdez, C. Rodríguez, M. E. Vázquez-Méndez and A. Martínez (Universidad de Santiago de Compostela)

> *Mathematical modelling and optimal control methods in waste water discharges*

J. I. Díaz (Univ. Complutense de Madrid) and J. L. Lions (Collège de France)

> *On the approximate controllability of Stackelberg–Nash strategies*

F. J. Elorza (Universidad Politécnica de Madrid)

> *Transport of pollutants in ground water and low permeability rocks*

J. Macías, C. Parés and M. J. Castro (Universidad de Málaga)

> *Numerical simulation in Oceanography. Applications to the Alboran Sea and the Strait of Gibraltar*

M. W. Saaltink and J. Carrera (Univ. Politècnica de Catalunya, Barcelona)

> *Simulation of reactive transport in groundwater. A comparison of two calculation methods*

G. Winter, J. Betancor and G. Montero (Universidad de Las Palmas)

> *3D Simulation in the lower troposphere: wind field adjustment to observational data and dispersion of air pollutants from combustion of sulfur-containing fuel*

Venice

Organiser: Elio Canestrelli

A. Bergamasco (Ist. per lo Studio della Dinamica delle Grandi Masse, Venice)

> *Coupling physical and biological modelling in coastal ecosystems: the Venice lagoon example*

V. Casulli (Università di Trento)

> *A mathematical model of the Venice lagoon*

G. Gambolati (Università degli Studi, Padova)

> *The mathematical model of the Venice subsurface system*

A. Quarteroni (Istituto Politecnico di Milano and CRS4 Cagliari)

> *Physical-numerical modelling of environmental processes*

Contents

Part A: Oceanic pollution control

A. BERMÚDEZ, C. RODRÍGUEZ, M. E. VÁZQUEZ-MÉNDEZ
AND A. MARTÍNEZ 3
    *Mathematical modelling and optimal control methods in waste
water discharges*

J. I. DÍAZ AND J. L. LIONS 17
    On the approximate controllability of Stackelberg–Nash strategies

G. WINTER, J. BETANCOR AND G. MONTERO 29
    *3D Simulation in the lower troposphere: wind field adjustment to
observational data and dispersion of air pollutants from combustion
of sulfur-containing fuel*

Part B: Numerical methods in oceanic circulation

R. BERMEJO 55
    *Eulerian versus semi-Lagrangian schemes in some ocean
circulation problems: a preliminary study*

J. MACÍAS, C. PARÉS AND M. J. CASTRO 75
    *Numerical simulation in Oceanography. Applications to the
Alboran Sea and the Strait of Gibraltar*

M. W. SAALTINK AND J. CARRERA 99
    *Simulation of reactive transport in groundwater. A comparison of
two calculation methods*

Oceanic pollution control

Mathematical modelling and optimal control methods in waste water discharges

A. Bermúdez[1], C. Rodríguez[1], M. E. Vázquez-Méndez[1], and A. Martínez[2]

[1] Departamento de Matemática Aplicada, Facultad de Matemáticas,
Universidad de Santiago de Compostela, 15706 Santiago, Spain
bermudez@zmat.usc.es, carmen@zmat.usc.es, ernesto@lugo.usc.es
[2] Departamento de Matemática Aplicada, E.T.S.I. Telecomunicaciones,
Universidad de Vigo, 36200 Vigo, Spain
aurea@dma.uvigo.es

1 Introduction

In this work we show how mathematical models and optimal control techniques can help to solve certain problems of environmental engineering —more precisely, water pollution problems arising from discharges into coastal areas or rivers.

Usually, waste waters originated from urban areas or industry undergo a physico-chemical and biological treatment in a plant. Then they are discharged through an outfall into an aquatic medium like a lake, a river or a coastal area, at an adequate distance from protected areas.

Before building such a system, studies of environmental impact are necessary in order to ensure that pollution does not reach swimming areas or marine culture areas. At this stage, mathematical models can be very useful, because they are cheaper and less aggressive than experimental methods. Furthermore, the answer can be obtained in a shorter time (see for instance [12], [24], [3], [13], [5], [6], [7]).

There are two main types of models, corresponding to the two following stages of effluent flow:

– Buoyant flow from the point of discharge towards the surface (jet models).
– Horizontal transport by current action from the final level in the previous stage (farfield models).

The first ones are systems of ordinary differential equations along the axis of the jet, while farfield models involve partial differential equations. In the present paper we deal with the second ones.

Very often, treatment plants discharge waste waters through outfalls which are placed in the same area (estuary, lake, etc.). Thus, all of them contribute to water pollution. In these circumstances, the problem of design and management of the whole sytem of treatment plants and outfalls arises. Optimization methods can help decision makers in formulating rational policies in order to minimize costs while keeping the prescribed levels of water quality (see for instance [14], [20], [4], [8], [11], [7]). An example is considered in Section 3. The problem is formulated as a pointwise optimal control problem with state and control constraints (see [21], [22], [23]). We present numerical results for a real problem posed in the *ría* of Vigo (Spain).

2   Mathematical models

First of all, it is convenient to notice that, since the volume of discharges is small compared with that of receiving waters, hydrodynamical equations may decouple from pollution dispersion equations. Therefore, the first step is to set and solve a model for simulating flows in the area under consideration. Then mathematical models can be used to simulate the dispersion of pollutants. As an importat example, we will present a system of partial differential equations governing the evolution of the Biological Oxigen Demand (BOD) and the Dissolved Oxigen (DO).

2.1   Hydrodynamic models

Currents are very often the main factor for dispersion of pollutants in farfield. In this section we recall the Saint Venant equations, which yield a useful mathematical model for hydrodynamic flows in shallow regions.

Consider an incompressible viscous newtonian fluid in a shallow domain defined as follows (see figure 1):

$$\Lambda(t) = \{(x_1, x_2, x_3) : (x_1, x_2) \in \Omega, \; b(x_1, x_2) \leq x_3 \leq b(x_1, x_2) + h(x_1, x_2, t)\} \quad (1)$$

where:

- Ω is the x_1, x_2 projection of the domain filled by the fluid,
- $h(x_1, x_2, t)$ is the height of the fluid layer at a point (x_1, x_2) and at a time t,
- $x_3 = b(x_1, x_2)$ is the equation of the bottom surface,
- $H(x_1, x_2) = A - b(x_1, x_2)$ is the depth from a fixed reference level A,
- $\eta(x_1, x_2, t) = h(x_1, x_2, t) - H(x_1, x_2)$ the surface elevation from the reference level A.

Assuming that pressure is hydrostatic and integrating the incompressible Navier–Stokes equations, the following system of partial differential equations can be obtained:

1. Mass conservation equation:

$$\frac{\partial h}{\partial t} + \frac{\partial (hu_1)}{\partial x_1} + \frac{\partial (hu_2)}{\partial x_2} = 0. \quad (2)$$

2. Momentum conservation equations:

$$\frac{\partial (hu_1)}{\partial t} + \frac{\partial (hu_1^2)}{\partial x_1} + \frac{\partial (hu_1 u_2)}{\partial x_2} + \frac{\partial}{\partial x_1} \int_b^{b+h} \hat{U}_1^{\,2} \, dx_3$$

$$+ \frac{\partial}{\partial x_2} \int_b^{b+h} \hat{U}_1 \hat{U}_2 \, dx_3 + \frac{\partial p_a}{\partial x_1} + gh\frac{\partial \eta}{\partial x_1} = 2\omega \sin \Phi hu_2$$

$$+ \frac{1}{\rho}\gamma_{10}v_1 \mid v \mid - \frac{gu_1\sqrt{u_1^2 + u_2^2}}{C^2}, \quad (3)$$

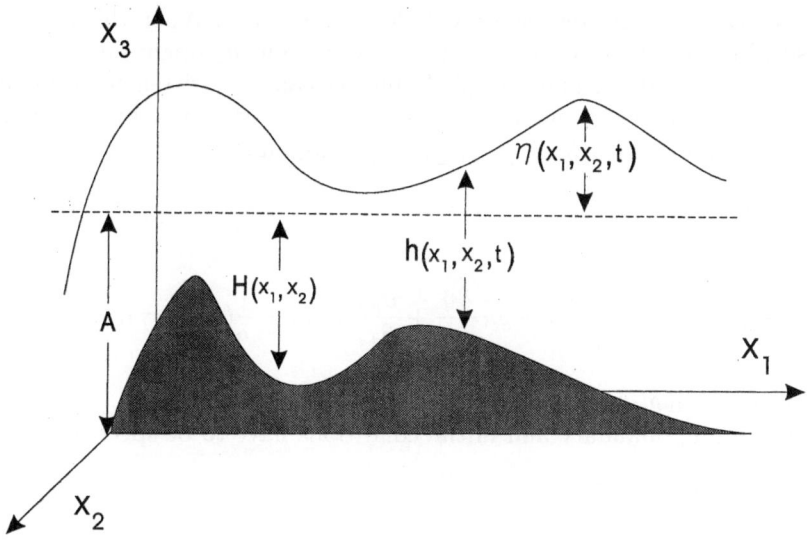

Fig. 1. The shallow domain.

$$\frac{\partial(hu_2)}{\partial t} + \frac{\partial(hu_1u_2)}{\partial x_1} + \frac{\partial(hu_2^2)}{\partial x_2} + \frac{\partial}{\partial x_1}\int_b^{b+h}\hat{U}_1\hat{U}_2\,dx_3$$

$$+\frac{\partial}{\partial x_2}\int_b^{b+h}\hat{U}_2^{\ 2}\,dx_3 + \frac{\partial p_a}{\partial x_2} + gh\frac{\partial\eta}{\partial x_2} = -2\omega\sin\Phi hu_1$$

$$+\frac{1}{\rho}\gamma_{10}v_2\mid v\mid -\frac{gu_2\sqrt{u_1^2+u_2^2}}{C^2}, \tag{4}$$

where:

- u_1 and u_2, the horizontal mean velocities, are defined by $u_1 = \frac{1}{h}\int_b^{b+h}U_1\,dx_3$, $u_2 = \frac{1}{h}\int_b^{b+h}U_2\,dx_3$,
- $\hat{U}_1 = U_1 - u_1$, $\hat{U}_2 = U_2 - u_2$,
- g is gravity,
- ρ is density,
- p_a is atmospheric pressure,
- v is the velocity of wind 10 m above water surface,
- ω is the angular velocity of the earth,
- C is the Chezy coefficient,
- Φ is the north latitude,
- $\gamma_{10}v|v| = \tau_w$ (wind stress),
- $\rho gu|u|/C^2 = \tau_f$ (bottom friction stress).

The four terms involving $\hat{U}_1$ and $\hat{U}_2$ are called the Reynolds stresses and represent the dispersive effects due to velocity fluctuations from the mean. Sometimes they can be neglected, as we will do in this paper. When this is not done, a "closure" giving these terms as functions of u_1 and u_2 might be used.

The shallow water equations (2, 3 and 4) can be written in terms of the conservative variables η and Q (the flow rate vector, given by $Q_1 = u_1 h = \int_b^{b+h} U_1 \, dx_3$ and $Q_2 = u_2 h = \int_b^{b+h} U_2 \, dx_3$), as follows:

$$
\begin{cases}
\dfrac{\partial \eta}{\partial t} + \nabla \cdot Q = 0 \\[2mm]
\dfrac{\partial Q_i}{\partial t} + \displaystyle\sum_{j=1}^{2} \left(\dfrac{\partial (u_j Q_i)}{\partial x_j} \right) + gh \dfrac{\partial \eta}{\partial x_i} + \dfrac{\partial p_a}{\partial x_i} = F_i + \dfrac{1}{\rho}(\tau_w - \tau_f), \quad i = 1, 2,
\end{cases}
\tag{5}
$$

where $F = 2\omega \, \mathrm{sen}\Phi(Q_2, -Q_1)$ accounts for the Coriolis effect.

Furthermore, boundary and initial conditions have to be specified:

(i) Coast or effluent (Γ_0):

$$
Q \cdot \nu = f, \quad \nu \text{ unit normal vector } (f = 0 \text{ on the coast}). \tag{6}
$$

(ii) Open sea (Γ_1):
$$
h = \phi + H \quad (\phi \text{ a given function}). \tag{7}
$$

(iii) Initial conditions:
$$
h(x, 0) = h_0(x) \tag{8}
$$
$$
u(x, 0) = u_0(x) \tag{9}
$$

2.2   Pollutant dispersion: the BOD/DO model

In order to control water pollution, some parameters are used which indicate the quality level of liquid media and their capacity to hold aquatic life. Among these indicators are dissolved oxygen, heavy metals, temperature, pH, radioactivity, etc.

Oxygen plays a major role in all kinds of life. In particular, it is used by bacterias to decompose organic matter. If the oxygen demand is not satisfied, plancton and other higher forms of animal life disappear. However, the decomposition of organic matter goes on by anaerobic processes which do not use oxygen but produce sulfure of hydrogen and methane, both having nauseous smell. The level of dissolved oxygen depends on two categories of factors:

1. The quantity of organic matter in the water to be decomposed and the mechanism for this decomposition.
2. Some physical conditions as temperature, depth, flow rate, turbulence, etc.

The organic matter can be measured in terms of the oxygen needed to decompose it. This is the so-called *biological oxygen demand* (BOD). If the pollution level is not too high, this demand can be satisfied by the dissolved oxygen. Notice that the dissolved oxygen is very sensitive to wastewater discharges, namely to the thermal ones. Indeed, at high temperatures the solubily of oxygen decreases while activity of microorganisms —which is oxygen consuming— increases.

Let us denote by ρ_1 (respectively ρ_2) the density of BOD (respectively DO). Then they satisfy the following system of partial differential equations (see [7], [30]):

$$\frac{\partial(\rho_1 h)}{\partial t} + \nabla \cdot (\rho_1 h u) - \beta_1 \Delta \rho_1 = -\kappa_1 \rho_1 h + \sum_{j=1}^{N_E} m_j \delta(x - P_j), \qquad (10)$$

$$\frac{\partial(\rho_2 h)}{\partial t} + \nabla \cdot (\rho_2 h u) - \beta_2 \Delta \rho_2 = -\kappa_1 \rho_1 h + \kappa_2(d_s - \rho_2)$$

$$-r_P M + \frac{I_B}{a + b I_B + c I_B^2} M, \qquad (11)$$

where:

- h is the height of water,
- u is the horizontal mean velocity,
- N_E is the number of outfalls,
- m_j is the discharge of BOD at a point P_j,
- $\delta(x - P_j)$ denotes the Dirac measure at a point P_j,
- β_1 and β_2 are horizontal viscosity coefficients,
- κ_1 is a kinetic parameter related to temperature,
- κ_2 is the interface transfer rate for oxygen,
- d_s is the oxygen saturation density in water,
- I_B is the intensity of sunlight on the bottom,
- M is the surface population density of algae,
- r_p is the respiration coefficient of algae,
- a, b and c are empirical coefficients.

3 Application of optimal control methods: Optimal management of a wastewater treatment system

3.1 Statement of the problem

In this section we suppose that several (N_E) treatment plants discharge waste waters through outfalls which are placed in the same domain $\Omega \subset \mathbb{R}^2$ occupied by shallow waters (estuary, lake, etc.) and, thus, all of them contribute to water pollution. Moreover, each plant has an associated cost ot treatment (f_j) which is a function of the final pollutant (BOD) concentrations. These functions can be different from one to another by several reasons, such as the technology, the nature of pollution content of arriving waste waters, etc.

Finally, we assume that there are N_Z areas in Ω (beaches, fish nurseries, etc.), where the pollution level has to be below the maximun value permitted by official regulations.

Then the following problem arises: What is the reduction of pollutant concentration to be made at each plant in order to minimize the total cost of the whole system, while keeping the prescribed levels of water quality in the protected areas?

Taking BOD/DO as water quality indicators and assuming that the depuration cost in a particular depuration plant is a function of the BOD discharged at the end of the process through the corresponding outfall, the optimal control problem can be written as follows:

Find the values (after depuration) of BOD concentrations $m_j(t) \geq 0$, $j = 1, \ldots, N_E$, which verify the state system (non-conservative form of the model (10)–(11) without algae effects and with boundary and initial conditions):

$$\left.\begin{array}{c} \dfrac{\partial \rho_1}{\partial t} + u\nabla\rho_1 - \beta_1\Delta\rho_1 = -\kappa_1\rho_1 + \dfrac{1}{h}\sum_{j=1}^{N_E} m_j\delta(x - P_j) \ \text{in} \ \Omega \times (0,T) \\[3mm] \dfrac{\partial\rho_1}{\partial n} = 0 \ \text{on} \ \Gamma \times (0,T) \\[3mm] \rho_1(x,0) = 0 \ \text{in} \ \Omega; \\[5mm] \dfrac{\partial\rho_2}{\partial t} + u\nabla\rho_2 - \beta_2\Delta\rho_2 = -\kappa_1\rho_1 + \dfrac{1}{h}\kappa_2(d_s - \rho_2) \ \text{in} \ \Omega \times (0,T) \\[3mm] \dfrac{\partial\rho_2}{\partial n} = 0 \ \text{on} \ \Gamma \times (0,T) \\[3mm] \rho_2(x,0) = \rho_{20}(x) \ \text{in} \ \Omega; \end{array}\right\} \quad (12)$$

satisfy the constraints

$$\rho_{1|A_i\times(0,T)} \leq \sigma_i, \quad i = 1, \ldots, N_Z, \tag{13}$$

$$\rho_{2|A_i\times(0,T)} \geq \zeta_i, \quad i = 1, \ldots, N_Z; \tag{14}$$

and minimize the cost function

$$J(m) = \sum_{j=1}^{N_E} \int_0^T f_j\left(m_j(t)\right) dt. \tag{15}$$

3.2   Discretization

The state equations are discretized by using finite element and characteristic methods (see [30]). Then we get a numerical approximation of the state variables (BOD and DO) at some grid points and time steps $(\rho_{ij}^n(x) \approx \rho_i(x_j, t_n))$, and we define the function g, which is putting together the discretized constraints:

$$g: \mathbb{R}^{N \times N_E} \longrightarrow \mathbb{R}^{N \times N_{VZ}} \times \mathbb{R}^{N \times N_{VZ}} \times \mathbb{R}^{N \times N_E}$$
$$m \longmapsto g(m) = (\underbrace{\rho_1 - \sigma, \zeta - \rho_2}_{=g_1(m)}, \underbrace{-m}_{=g_2(m)})^t, \tag{16}$$

where:

- m is the vector consisting of all of the discharges at all times,
- N_{VZ} is the number of vertices in the protected areas,
- ρ_i is a vector of values of ρ_i at vertices included in the protected areas and for all times.

The cost functional is also discretized by using a quadrature formula, and we define the new cost:

$$\hat{J} \colon \mathbb{R}^{N \times N_E} \longrightarrow \mathbb{R}$$

$$m \qquad \longmapsto \hat{J}(m) = \Delta t \sum_{j=1}^{N_E} \sum_{n=0}^{N-1} Q_{jn} f_j(m_{jn}), \qquad (17)$$

where m_{jn} is the amount of BOD discharged in P_j at a time t_n, and Q_{jn} are the weights of the quadrature formula.

Then the *discrete* optimal control problem has the following form:

$$(\mathcal{P}_{\mathcal{F}}) \quad \left\{ \begin{array}{c} \min_{m \in \mathbb{R}^{N \times N_E}} \hat{J}(m) \\ \text{such that } g(m) \leq 0. \end{array} \right.$$

3.3 Solving the discrete optimization problem: an interior point method

We have solved the problem $(\mathcal{P}_{\mathcal{F}})$ by means of an admissible points method which is based on a globally convergent algorithm introduced by Herskovits [17] and Panier *et al* [25] for nonlinear constraints.

If we denote the vector of the dual variables by (λ, θ), then we can write the first order Karush–Kuhn–Tucker optimality conditions for our problem as follows:

$$\nabla \hat{J}(m) + \nabla g_1 \lambda - I\theta = 0, \qquad (18)$$

$$G_1(m)\lambda = 0, \qquad G_2(m)\theta = 0, \qquad (19)$$

$$\lambda \geq 0, \qquad \theta \geq 0, \qquad (20)$$

$$g_1(m) \leq 0, \qquad g_2(m) \leq 0, \qquad (21)$$

where $G_1(m)$ and $G_2(m)$ are diagonal matrices, with $((G_i)(m))_{jj} = (g_i(m))_j$.

The basic idea of the algorithm of admissible points consists of solving the system of equations (18)–(19) in (m, λ, θ) by using an iterative method, in such a way that the conditions (20)–(21) hold at each iteration.

For a given point $(m^k, \lambda^k, \theta^k)^t$, the Newton method applied to the previous system computes the next iteration $(m_0^{k+1}, \lambda_0^{k+1}, \theta_0^{k+1})^t$ by solving

$$\begin{pmatrix} m_0^{k+1} \\ \lambda_0^{k+1} \\ \theta_0^{k+1} \end{pmatrix} = \begin{pmatrix} m^k \\ \lambda^k \\ \theta^k \end{pmatrix} - \begin{pmatrix} H(m^k, \lambda^k, \theta^k) & \nabla g_1 & -I \\ \Lambda^k(\nabla g_1)^t & G_1^k & 0 \\ -\Theta^k & 0 & G_2^k \end{pmatrix}^{-1} \begin{pmatrix} \nabla \hat{J}(m^k) + \nabla g_1 \lambda^k - I\theta^k \\ G_1^k \lambda^k \\ G_2^k \theta^k \end{pmatrix}$$

where:

- $H(m, \lambda, \theta) = \nabla^2 \hat{J}(m) + \sum_{i=1}^{q} \lambda_i \nabla^2 g_{1i}(m) + \sum_{i=1}^{p} \theta_i \nabla^2 g_{2i}(m)$ is the hessian of the lagrangian,
- $G_1^k = G_1(m^k)$, $G_2^k = G_2(m^k)$,
- Λ^k, Θ^k are diagonal matrices, with $(\Lambda^k)_{ii} = (\lambda^k)_i$ and $(\Theta^k)_{ii} = (\theta^k)_i$.

In general, the point $(m_0^{k+1}, \lambda_0^{k+1}, \theta_0^{k+1})^t$ is not feasible (since the equations (20) and (21) do not hold). Then, we define $d^k = m_0^{k+1} - m^k$ as a search direction in m and rewrite the previous equality by computing $(d^k, \lambda_0^{k+1}, \theta_0^{k+1})^t$ as the solution of the following linear system:

$$
\begin{pmatrix} H(m^k, \lambda^k, \theta^k) & \nabla g_1 & -I \\ \Lambda^k (\nabla g_1)^t & G_1^k & 0 \\ -\Theta^k & 0 & G_2^k \end{pmatrix} \begin{pmatrix} d^k \\ \lambda_0^{k+1} \\ \theta_0^{k+1} \end{pmatrix} = \begin{pmatrix} -\nabla \hat{J}(m^k) \\ 0 \\ 0 \end{pmatrix} \qquad (22)
$$

Now, in order to determine the new primal point m^{k+1}, we perform a line search along d^k to obtain a step t^k which leads us to a new admissible point $m^{k+1} = m^k + t^k d^k$ where the cost reduction is satisfactory. Finally, the new value of the dual variable $(\lambda^{k+1}, \theta^{k+1})^t$ can be computed from $(\lambda_0^{k+1}, \theta_0^{k+1})^t$ by several updating methods.

According to this, the general sketch of the algorithm is the following:

ADMISSIBLE POINTS ALGORITHM	
Previous Information	1. Compute $g_1(0)$, ∇g_1 2. Choose $(m^0, \lambda^0, \theta^0)^t$ such that $g(m^0) \leq 0$, $\lambda^0 \geq 0$, $\theta^0 \geq 0$
STEP 1	Compute the descent direction d^k by solving the linear system (22)
STEP 2	Compute the step length t^k by employing a line search technique and define $m^{k+1} = m^k + t^k d^k$
STEP 3	Update the dual variable: Define $(\lambda^{k+1}, \theta^{k+1})$ from $(\lambda_0^{k+1}, \theta_0^{k+1})$
STEP 4	Test of Convergence: i. If it is OK $\longrightarrow$ stop algorithm and accept m^{k+1} as solution of the problem $(\mathcal{P}_{\mathcal{F}})$ ii. If it is not OK $\longrightarrow$ go back to STEP 1

3.4   Numerical results

The problem $(\mathcal{P}_{\mathcal{F}})$ has been solved for a study case corresponding to the Ría of Vigo (Spain). First of all, we have calculated the velocity and the height of the water by solving the shallow water equations (5) (figure 2 shows the velocity field at high tide). Then we have considered two discharge points ($N_E = 2$) as well as two protected areas ($N_Z = 2$), and we assume that the pollution level in area 1 must be lower than in area 2:

maximum BOD	*minimum DO*
$\sigma_1 = 5.810^{-2}$ Kg/m^3,	$\delta_1 = 7.186410^{-3}$ Kg/m^3
$\sigma_2 = 6.610^{-2}$ Kg/m^3,	$\delta_2 = 7.035410^{-3}$ Kg/m^3

Moreover, we suppose that the cost of the depuration is the same for the two purifying plants, and that the mass flow rate of BOD arriving to both is 150 Kg/s, so the cost function above this value is constant (see figure 5).

In the figure 3 we show the isolines for concentration of BOD at high tide. State constraints hold everywhere in the protected areas and saturate at one vertex in zone 1. At low tide, after a tidal cycle, the BOD concentrations can be seen in the figure 4. Now saturation takes place at one vertex of area 2.

The optimal values of discharges are given in the figure 6. One can observe that during rising tide the discharge rate is greater at point 2 than at point 1. However, during ebb tide (after $t = 60$) the flow rate decreases at P_2 and increases at P_1. This is an obvious consequence of the position of the two outfalls.

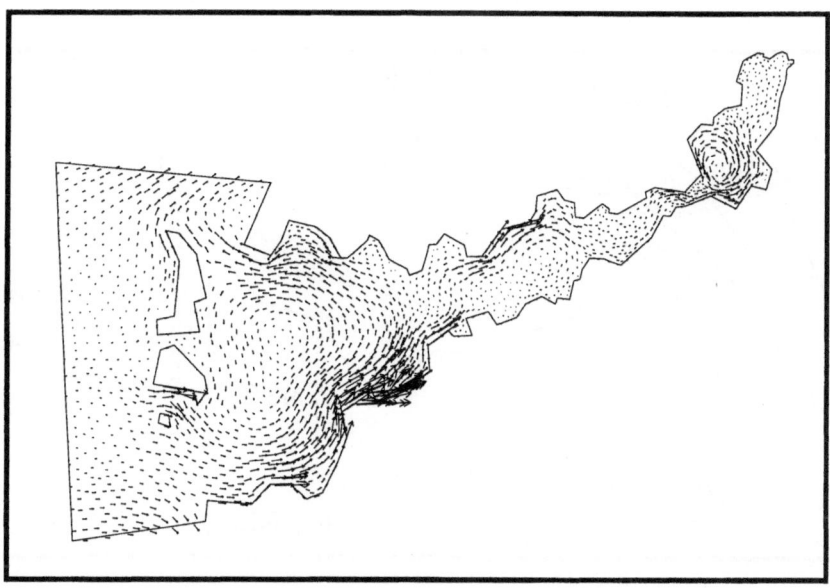

Fig. 2. The velocity field at high tide.

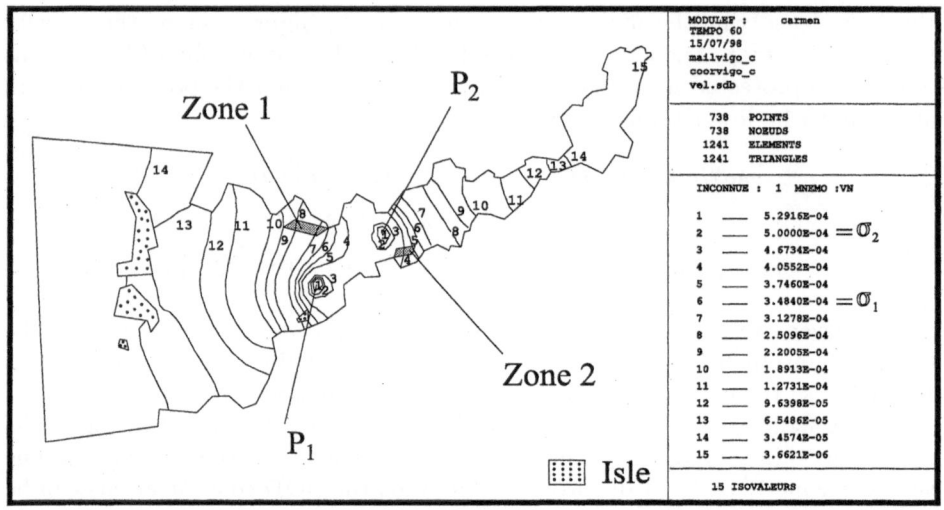

Fig. 3. BOD concentration at high tide.

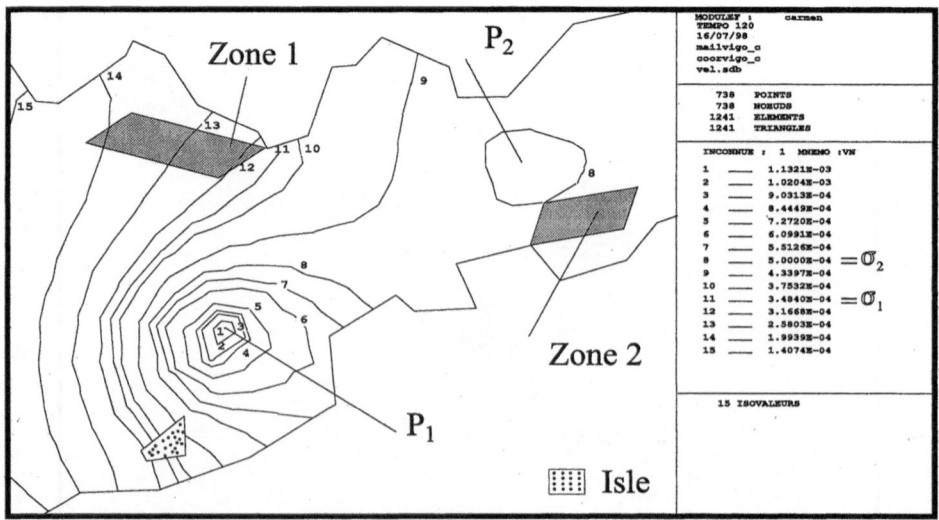

Fig. 4. BOD concentration at low tide.

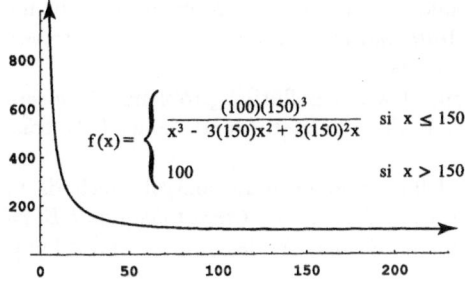

$$f(x) = \begin{cases} \dfrac{(100)(150)^3}{x^3 - 3(150)x^2 + 3(150)^2 x} & \text{si } x \le 150 \\[2mm] 100 & \text{si } x > 150 \end{cases}$$

Fig. 5. Cost function.

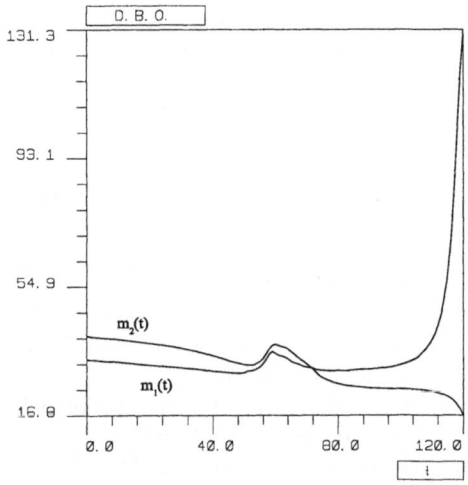

Fig. 6. Optimal discharges during a tidal cycle.

References

1. Abbot [1985]: *Computational Hydraulics,* Pitman, Boston.
2. Alcrudo, F., García-Navarro, P. and Savirón, J. M. [1993]: Flux-difference splitting for 1D open channel flow equations, *Int. J. Num. Methods in Eng.* **14**, 1009–1018.
3. Ames, W. F. [1988]: Analysis of mathematical models for pollutant transport and dissipation, *Comput. Math. with Appl.* **16**, 939–985.
4. Antonios, M. N. [1989]: Optimization problems related to water quality control in aquatic ecosystems, *Comput. Math. with Appl.* **18**, 851–870.

5. Bermúdez, A. [1993]: Mathematical techniques for some environmental problems related to water pollution control, in *Mathematics, Climate and Environment*, Díaz, J. I. and Lions, J. L. eds., Masson, Paris.

6. Bermúdez, A. [1994]: Numerical modelling of water pollution problems, *Environment, Economics and their Mathematical Models*, Díaz, J. I. and Lions, J. L. eds., Masson, Paris.

7. Bermúdez, A. [1997]: Mathematical modelling and optimal control methods in water pollution problems, *The Mathematics of Models for Climatology and Environment, Nato ASI Series I 48*, Díaz, J. I. ed., Springer Verlag, Berlin, Heidelberg, New York.

8. Bermúdez, A., Martínez, A., Rodríguez, C. [1991]: Un problème de contrôle ponctuel lié à l'emplacement optimal d'emissaires d'évacuation sous-marins, *C. R. Acad. Sci. Paris* t. **313**, Série I, 515–518.

9. Bermúdez, A., Rodríguez, C., Vilar, M. A. [1991]: Solving shallow water equations by a mixed implicit finite element method, *IMA J. of Num. Analysis* **11**, 79–97.

10. Bermúdez, A., Vázquez, M. E. [1994]: Upwind methods for hyperbolic conservation laws with source terms, *Computers and Fluids* **23**, n. 8 1049–1071.

11. Bogobowicz, A. [1991]: Theoretical aspects of modeling and control of water quality in river section, *Appl. Math. and Comp.* **41**, 35–60.

12. Brebbia, C. A. (Ed.) [1976]: *Mathematical Model for Environmental Problems*, Pentech Press, London.

13. Gambolati, G., Rinaldo, A., Brebbia, C. A., Gray, W. G., Pinder, G. F. [1990]: *Computational Methods in Surface Hydrology*, Springer Verlag, Berlin.

14. Haimes, Y. Y. [1976]: Hierarchical analysis of water ressources systems, McGraw Hill, New York.

15. Herskovits, J., Santos, G. [1982]: A two-stage feasible direction algorithm including variable metric techniques for nonlinear optimization problems, *Rapports de Recherche*, INRIA.

16. Herskovits, J. [1992]: An interior point technique for nonlinear optimization, *Rapports de Recherche*, INRIA.

17. Herskovits, J., Santos, G. [1997]: On the computer implemetation of feasible Direction Point Algorithms for nonlinear optimization. COPPE - federal University of Rio de Janeiro, Mechanical Engineering Program, Caixa Postal 68503, 21945-970, Rio de Janeiro, Brazil.

18. Lions, J. L. [1968]: *Contrôle Optimal des Systèmes Gouvernés par des Équations aux Dérivées Partielles*, Dunod, Paris.

19. Lions, J. L. [1979]: Nouveaux espaces fonctionnels en théorie du contrôle des systèmes distribués, *C. R. Acad. Sci. Paris* t. **289**, Série I, 315–319.

20. Loucks, D. P., Stedinger, J. R., Haith, D. A. [1978]: *Water Resources Systems Planning and Analysis*, Prentice Hall, New York.

21. Martínez, A., Rodríguez, C., Vázquez Méndez, M. E. [1998]: Resolución numérica de un problema de control relativo a la depuración de aguas residuales, Actas del XV CEDYA/X CMA, Universidad de Vigo, Spain.

22. Martínez, A., Rodríguez, C., Vázquez-Méndez, M. E. [1998]: *Theoretical and numerical analysis of an optimal control problem related to wastewater treatment*, Preprint, Dept. Matemática Aplicada, Univ. Santiago de Compostela.

23. Martínez, A., Rodríguez, C., Vázquez-Méndez, M. E. [1998]: *A control problem related to wastewater treatment*. C. R. Acad. Sci. Paris, Serie I, in press.

24. Nihoul, J. (Ed.) [1975]: *Modelling of Marine Systems*, Elsevier, Amsterdam.

25. Panier, E. R., Tits, A.-L., Herskovits, J. [1988]: A QP-free, globally convergent, locally superlinearly convergent algorithm for inequality constrained optimization, *SIAM Journal of Control and Optimization* **26**, 788–810.

26. Quetin, B., De Rouville, M. [1986]: Submarine sewer outfalls —A design manual, *Marine Pollution Bulletin* **17**, 133–183.

27. Rahman, M. [1988]: *The Hydrodynamics of Waves and Tides, with Applications*, Computational Mechanics Publications, Southampton.

28. Stoker, J. J. [1957]: *Water Waves*, Interscience, New York.

29. Thomann, R. V. [1972]: *Systems Analysis and Water Quality Management*, Environmental Research and Applications Inc, New York.

30. Vázquez-Méndez, M. E. [1992]: *Contribución a la resolución numérica de modelos para el estudio de la contaminación de aguas*, Master Thesis, Publ. Depart. Matemática Aplicada, University of Santiago de Compostela.

On the approximate controllability of Stackelberg-Nash strategies

J. I. Díaz[1] and J. L. Lions[2]

[1] Departamento de Matemática Aplicada, Facultad de Matemáticas, Universidad Complutense de Madrid, 28040 Madrid, Spain, ji_diaz@mat.ucm.es
[2] Collège de France, 3 rue d'Ulm, 75231, Paris Cedex 05, France

1 Introduction

Let us consider a *distributed system,* i.e. a system whose state is defined by the solution of a Partial Differential Equation (PDE). We assume that we can *act* on this system by a *hierarchy of controls.* There is a "global" control v, which is the *leader,* and there are N "local" controls, denoted by $w_1, \ldots, w_N$, which are the *followers.* The followers, assuming that the leader has made a choice v of its policy, look for a *Nash equilibrium* of their cost functions (the criteria they are interested in). *Then* the leader makes its final choice for the whole system. *This is the Stackelberg–Nash strategy.*

Such situations arise in very many fields of Environment and of Engineering (and, by the way, for systems not necessarily described by PDE's). In order to explain more precisely our motivation, let us choose here an example taken from Environment: let us consider a resort lake, represented by a domain Ω of $\mathbb{R}^3$. The state of the system is denoted by $\mathbf{y}$. It is a vector function $\mathbf{y} = \{y_1, \ldots, y_N\}$, each y_i being a function of x and t, $x \in \Omega$, $t = $ time. The y_i's correspond to concentrations of various chemicals in the lake Ω or of living organisms. The y_i's are therefore given by the solution of a set of *diffusion equations.* In the resort, there are *local agents* or *local plants,* $P_1, \ldots, P_N$. Each plant P_i can decide (with some constraints) its policy w_i. There is also a general manager of the resort. He (or she) has the choice of the policy denoted by v. Therefore the *state equations* are given by

$$\frac{\partial \mathbf{y}}{\partial t} + \mathcal{A}(\mathbf{y}) = \text{sources} + \text{sinks} + \text{global control } v + \text{local control } \{w_1, \ldots, w_N\}, \quad (1)$$

where the *initial state* is supposed to be given,

$$\mathbf{y}(x, 0) = \mathbf{y}_0(x), \quad (2)$$

and where there are appropriate boundary conditions (of course this is made more precise in the next section of this paper). The general goal of the manager v is to maintain the lake as "clean" as possible. In other words, if the situation at $t = 0$ is not entirely satisfactory, he (or she) wants to "drive the system" at a chosen time horizon T as *close as possible to an ideal state,* denoted by $\mathbf{y}^T$. Each plant P_i has essentially the same goal, but of course, P_i will be particularly careful to the state **y** *near its location*. Let ρ_i be a smooth function given in $\overline{\Omega}$ such that

$$\rho_i(x) \geq 0, \ \rho_i = 1 \text{ near the location of } P_i. \tag{3}$$

Then P_i will try to choose w_i such that the state at time T, $\mathbf{y}(x, T)$, be "close" to $\rho_i \mathbf{y}^T$, and to achieve this *at minimum cost*. This leads to the introduction of

$$J_i(v; w_1, \ldots, w_N) = \frac{1}{2} \||w_i\||^2 + \frac{\alpha_i}{2} \left\| \rho_i(\mathbf{y}(., T) - \mathbf{y}^T) \right\|^2, \tag{4}$$

where $\||w_i\||$ represents the cost of w_i, α_i is a given positive constant and $\left\| \rho_i(\mathbf{y}(., T) - \mathbf{y}^T) \right\|$ is a measure of the "localized distance" between the actual state at time T and the desired state $\mathbf{y}^T$.

Remark 1.1 We have assumed here that the system (1), (2) (together with appropriate boundary conditions) admits a unique solution $\mathbf{y}(x, t; v; w_1, \ldots, w_N)$. In (4), $\mathbf{y}(., T)$ denotes the function $x \mapsto \mathbf{y}(x, T; v; w_1, \ldots, w_N)$.

The "local" controls $w_1, \ldots, w_N$ assume that *the leader has made a choice* v and they try to find a *Nash equilibrium* of their cost J_i, i.e. they look for $w_1, \ldots, w_N$ (*as functions of* v) such that

$$J_i(v; w_1, \ldots, w_{i-1}, w_i, w_{i+1}, \ldots, w_N) \leq J_i(v; w_1, \ldots, w_{i-1}, \widehat{w_i}, w_{i+1}, \ldots, w_N),$$

for all $\widehat{w_i}$, for $i = 1, \ldots, N$. $\tag{5}$

If $\mathbf{w} = \{w_1, \ldots, w_N\}$ satisfies (5), one says it is a *Nash equilibrium*.

The leader v wants now that the *global* state (i.e. the state $\mathbf{y}(., T)$ in the whole domain Ω) to be *as close as possible* to $\mathbf{y}^T$. This will be possible, for any given function $\mathbf{y}^T$, if the problem is *approximately controllable*, i.e. if

$$\mathbf{y}(x, t; v; w_1, \ldots, w_N) \text{ describes a dense subset of the given state} \atop \text{space when } v \text{ spans the set of all controls available to the leader.} \tag{6}$$

Remark 1.2 We emphasize again that in (6) the controls w_i are chosen so that (5) is satisfied. Therefore they are functions of v.

Remark 1.3 The above strategy is of the *Stackelberg's type*. This strategy has been introduced by Stackelberg [12] in 1934 for problems arising in Economics. It has been used in problems of distributed systems in Lions [7], without reference to controllability questions and in Lions [8] in a different setting *without using Nash equilibria*.

Remark 1.4 We have explained the family of problems we are interested in for environment questions, but problems of this type arise in many other questions, such as the control of large engineering systems.

Remark 1.5 It is clear that $\mathbf{y}^T$ is *not* going to be an *arbitrary* function in the state space. Therefore the resort could be maintained in a satisfactory state even *without* the system being approximately controllable (in the sense of (6)). But if there is a serious degradation following, for instance, an accident, then the *initial* state can be "anything" so that it is certainly preferable to live in a "controllable resort"...

Remark 1.6 Of course, the *Stackelberg's type strategy* is not the only possible! One could also replace the *Nash equilibrium* by a *Pareto equilibrium* for the followers $w_1, \ldots, w_N$ (see, for instance, Lions [9]). Here all the controls w_i agree to work in a strategy where v is the leader, and they agree to work in the context of a *Nash equilibrium*. Their personal (selfish) interests are expressed in the cost functions J_i as we shall see in the next section.

Remark 1.7 In the above context *there does not always exist a Nash equilibrium*. We prove in Section 4 some sufficient conditions for the existence and uniqueness of a *Nash equilibrium*. We also present a general counterexample showing that those conditions are, in some sense, necessary. What we (essentially) show in this paper (the first of a series) is that for *linear* systems, if there is existence and uniqueness of a Nash equilibrium for the followers, then the leader *can control the system* (in the sense of approximate controllability). The study of the case of nonlinear systems is the main subject of Díaz and Lions [2].

The content of the rest of this paper is the following: In the next section we make precise the statement of our main result by taking *one* state equation, i.e. **y** is a *scalar* function y instead of a vector function $\{y_1, \ldots, y_N\}$. This is just for the sake of simplicity of the exposition. It is by no means a serious restriction. But we shall make a *very strong* assumption, namely that the state equation *is linear*. The proof of the approximate controllability will be given in Section 3. The study of suitable assumptions (and their optimality) implying the existence and uniqueness of a *Nash equilibrium* is carried out in Section 4. Finally, some further remarks are presented in Section 5.

2   Statement of the approximate controllability theorem

Let A be a second order elliptic operator in Ω :

$$A\varphi = -\sum_{i,j=1}^{N} \frac{\partial}{\partial x_i} \left(a_{i,j}(x) \frac{\partial \varphi}{\partial x_j} \right) + \sum_{i=1}^{N} a_i(x) \frac{\partial \varphi}{\partial x_i} + a_0(x)\varphi, \qquad (7)$$

where all coefficients are smooth enough and where

$$\sum_{i,j=1}^{N} a_{i,j}(x)\xi_i\xi_j \geq \alpha \sum_{i=1}^{N} \xi_i^2, \ \alpha > 0, \ x \in \overline{\Omega}. \qquad (8)$$

We assume that the state equation is given by

$$\frac{\partial y}{\partial t} + Ay = v\chi + \sum_{i=1}^{N} w_i\chi_i \qquad (9)$$

where

$$\chi \text{ is the characteristic function of } \mathcal{O} \subset \Omega, \text{ and}$$
$$\chi_i \text{ is the characteristic function of } \mathcal{O}_i \subset \Omega. \tag{10}$$

Remark 2.1 The control function $v(x,t)$ of the leader is distributed in $\mathcal{O}$ and the control function $w_i(x,t)$ of the follower "i" is distributed in $\mathcal{O}_i$.

Remark 2.2 All the results to follow are also valid for *boundary controls*. The case of distributed controls permits to avoid some difficulties of a purely technical type.

We assume that the *initial state* is

$$y(x,0) = 0, \ x \in \Omega. \tag{11}$$

Remark 2.3 Since the system is *linear*, there is no restriction in assuming the initial state to be zero, in the same way as there is no restriction in assuming in (9) that sources + sinks are zero (compare to (1)).

We assume that the *boundary conditions* are

$$y = 0 \ \text{ on } \ \partial\Omega \times (0,T). \tag{12}$$

Remark 2.4 Again (12) is not at all a serious restriction. We could consider as well y to be nonzero and that the following results apply for other boundary conditions.

We introduce now functions ρ_i such that

$$\left. \begin{array}{l} \rho_i \in L^\infty(\Omega), \ \rho_i \geq 0, \\ \rho_i = 1 \ \text{in a domain } \mathcal{G}_i \subset \Omega, \end{array} \right\} \tag{13}$$

and we define the cost function J_i (compare to (4))

$$J_i(v; w_1, \ldots, w_N) = \frac{1}{2} \int_0^T \int_{O_i} w_i^2 \, dx dt + \frac{\alpha_i}{2} \left\| \rho_i y(T; v, \mathbf{w}) - \rho_i y^{\mathbf{T}} \right\|^2, \tag{14}$$

where $\|\cdot\|$ is the norm in $L^2(\Omega)$.

Remark 2.5 In the case of the example presented in the Introduction, $\mathcal{G}_i$ is the region of the lake the plant P_i is particularly interested in (the place near P_i for instance!). If P_i is selfish, then $\rho_i = 0$ outside $\mathcal{G}_i$.

Remark 2.6 From a mathematical view point, the only hypothesis needed on ρ_i is that $\rho_i \in L^\infty(\Omega)$ (one could even take ρ_i in a suitable $L^p(\Omega)$ space, but this is irrelevant here).

Remark 2.7 We assume that

$$v \in L^2(\mathcal{O} \times (0,T)), \quad w_i \in L^2(\mathcal{O}_i \times (0,T))$$

and that $y(x,t;v,\mathbf{w})$ is *the* solution of (9), (11), (12).

Given $v \in L^2(\mathcal{O} \times (0,T))$, we now define (cf. (5))

$$\left. \begin{array}{c} \mathbf{w} = \{w_1, \ldots, w_N\}, \text{ a } \textit{Nash equilibrium} \text{ for the cost,} \\ \text{and functions } J_1, \ldots, J_N \text{ given by (14).} \end{array} \right\} \tag{15}$$

We will show in Section 3 how (under hypotheses which are presented in Section 4) that this *Nash equilibrium* can be defined as a function of v:

$$\mathbf{w} = \mathbf{w}(v) \text{ or } w_i = w_i(v), \ i = 1, \ldots, N. \tag{16}$$

We then replace in (9) w_i by $w_i(v)$:

$$\frac{\partial y}{\partial t} + Ay = v\chi + \sum_{i=1}^{N} w_i(v)\chi_i \tag{17}$$

subject to (11) and (12). The system (17), (11) and (12) admits a unique solution $y(x,t;v,\mathbf{w}(v))$. In Section 3 we prove the following result.

Theorem 2.1 *Assume that*

> *the set of inequalities (5) admits a unique solution (a Nash equilibrium).* (18)

Then, when v spans $L^2(\mathcal{O} \times (0,T))$, the functions $y(.,T;v,\mathbf{w}(v))$ describe a dense subset of $L^2(\Omega)$. In other words,

$$\begin{array}{c} \textit{there is approximate controllability of the system} \\ \textit{when a strategy of the Stackelberg–Nash type is followed.} \end{array} \tag{19}$$

3 Proof of the main theorem

3.1 Nash equilibrium

We have (5) iff

$$\int_0^T \int_{\mathcal{O}_i} w_i \widehat{w_i} \, dx dt + \alpha_i \int_\Omega \rho_i^2 (y(T;v,\mathbf{w}) - y^T) \, \widehat{y_i}(T) \, dx = 0, \ \forall \widehat{w_i}, \tag{20}$$

where $\widehat{y_i}$ is defined by

$$\left.\begin{array}{c}\dfrac{\partial \widehat{y_i}}{\partial t} + A\widehat{y_i} = \widehat{w_i}\chi_i,\\[2mm] \widehat{y_i}(0) = 0 \text{ in } \Omega, \ \widehat{y_i} = 0 \text{ in } \partial\Omega \times (0,T).\end{array}\right\} \tag{21}$$

In order to express (20) in a convenient form, we introduce the adjoint state p_i defined by

$$\left.\begin{array}{c}-\dfrac{\partial p_i}{\partial t} + A^* p_i = 0 \text{ in } \Omega \times (0,T),\\[2mm] p_i(x,T) = \rho_i^2(x)(y(x,T;v,\mathbf{w}) - y^T(x)) \text{ in } \Omega,\\[2mm] p_i = 0 \text{ in } \partial\Omega \times (0,T),\end{array}\right\} \tag{22}$$

where A^* stands for the adjoint of A. If we multiply (22) by $\widehat{y_i}$ and if we integrate by parts, we find

$$\int_\Omega \rho_i^2(y(T;v,\mathbf{w}) - y^T)\,\widehat{y_i}(T)\,dx = \int_0^T \int_\Omega p_i \widehat{w_i}\chi_i \, dx dt,$$

so that (20) becomes

$$\int_0^T \int_{O_i} (w_i + \alpha_i p_i)\widehat{w_i}\, dx dt = 0, \ \forall \widehat{w_i},$$

i.e.

$$w_i + \alpha_i p_i \chi_i = 0. \tag{23}$$

Then, if $\mathbf{w} = \{w_1, \ldots, w_N\}$ is a *Nash equilibrium*, we have

$$\left.\begin{array}{c}\dfrac{\partial y}{\partial t} + Ay + \displaystyle\sum_{i=1}^{N} \alpha_i p_i \chi_i = v\chi,\\[2mm] -\dfrac{\partial p_i}{\partial t} + A^* p_i = 0, \ i = 1, \ldots, N,\\[2mm] y(0) = 0, \ p_i(x,T) = \rho_i^2(x)(y(x,T;v,\mathbf{w}) - y^T(x)) \text{ in } \Omega,\\[2mm] y = 0, \ p_i = 0 \text{ in } \partial\Omega \times (0,T).\end{array}\right\} \tag{24}$$

We recall that here we are assuming the existence and uniqueness of a *Nash equilibrium* (hypothesis (18)). We return to that in Section 4.

3.2   Approximate controllability: Proof of Theorem 2.1

We want to show that the set described by $y(.,T;v)$ is dense in $L^2(\Omega)$, where y is the solution given by (24) and when v spans $L^2(\mathcal{O} \times (0,T))$. We do not restrict the problem by assuming that

$$y^T \equiv 0$$

(it suffices to use a translation argument). Let f be given in $L^2(\Omega)$ and let us assume that

$$(y(.,T;v), f) = 0, \ \forall v \in L^2(\Omega). \tag{25}$$

We want to show that $f \equiv 0$. Let us introduce the solution $\{\varphi, \psi_1, \ldots, \psi_N\}$ of the adjoint system

$$
\left.
\begin{aligned}
-\frac{\partial \varphi}{\partial t} + A^* \varphi &= 0, \\
\frac{\partial \psi_i}{\partial t} + A\psi_i &= -\alpha_i \varphi \chi_i, \\
\varphi(T) &= f + \sum_i \psi_i(T)\rho_i^2, \\
\psi_i(0) &= 0, \\
\varphi = 0, \ \psi_i &= 0 \text{ in } \partial\Omega \times (0,T).
\end{aligned}
\right\}
\tag{26}
$$

We multiply the first (resp. the second) equation in (26) by y (resp. p_i). We obtain

$$
\left.
\begin{aligned}
&-(f + \sum_i \psi_i(T)\rho_i^2, y(T)) + \int_0^T \int_\Omega \varphi\left(\frac{\partial y}{\partial t} + Ay\right) \, dx\,dt + \\
&\sum_i (\psi_i(T), p_i(T)) + \\
&+\sum_i \int_0^T \int_\Omega \psi_i\left(-\frac{\partial p_i}{\partial t} + A^* p_i\right) \, dx\,dt = -\sum_i \alpha_i \int_0^T \int_\Omega \varphi p_i \chi_i \, dx\,dt.
\end{aligned}
\right\}
\tag{27}
$$

Using (24) (where $y^T \equiv 0$), (27) reduces to

$$
-(f, y(T)) + \int_0^T \int_\Omega \varphi v \chi \, dx\,dt = 0.
\tag{28}
$$

Therefore, if (25) holds, then

$$
\varphi = 0 \text{ on } \mathcal{O} \times (0,T).
\tag{29}
$$

Using Mizohata's Uniqueness Theorem (see Mizohata [5] or Saut and Scheurer [10]) —this is the only place where some smoothness on the coefficients of A is needed— it follows from $(26)_1$ and (29) that

$$
\varphi = 0 \text{ on } \Omega \times (0,T).
\tag{30}
$$

Then $(26)_2$, $(26)_4$ and $\psi_i = 0$ in $\partial\Omega \times (0,T)$ imply that

$$
\psi_i = 0 \text{ in } \Omega \times (0,T), \ i = 1, \ldots, N,
\tag{31}
$$

so that $(26)_3$ gives $f \equiv 0$.

4   On the existence and uniqueness of Nash equilibrium

4.1   A criterion of existence and uniqueness

We consider the functionals (14). Let us define

$$
\left.
\begin{aligned}
\mathcal{H}_i &= L^2(\mathcal{O}_i \times (0,T)), \\
\mathcal{H} &= \prod_{i=1}^N \mathcal{H}_i, \\
L_i \widehat{w_i} &= \widehat{y_i}(T) \text{ (cf. (21))}, \text{ which defines } L_i \in L(\mathcal{H}_i; L^2(\Omega)).
\end{aligned}
\right\}
\tag{32}
$$

Since v is fixed, one can write

$$y(T; v, \mathbf{w}) = \sum_{i=1}^{N} L_i w_i + z^T, \; z^T \text{ fixed}. \tag{33}$$

With these notations (14) can be rewritten

$$J_i(v; \mathbf{w}) = \frac{1}{2} \|w_i\|_{\mathcal{H}_i}^2 + \frac{\alpha_i}{2} \left\| \rho_i \left(\sum_j L_j w_j - \eta^T \right) \right\|^2 \tag{34}$$

where $\eta^T = y^T - z^T$. Then $\mathbf{w} \in \mathcal{H}$ is a *Nash equilibrium* iff

$$(w_i, \widehat{w_i})_{\mathcal{H}_i} + \alpha_i \left(\rho_i \left(\sum_j L_j w_j - \eta^T \right), \; \rho_i L_i \widehat{w_i} \right) = 0, \; i = 1, \ldots, N, \; \forall \widehat{w_i}. \tag{35}$$

or

$$w_i + \alpha_i L_i^* \left(\rho_i^2 \sum_{j=1}^{N} L_j w_j \right) = \alpha_i L_i^* \left(\rho_i^2 \eta^T \right), \; i = 1, \ldots, N \tag{36}$$

(where $L_i^* \in \mathcal{L}(L^2(\Omega); \mathcal{H}_i)$ is the adjoint of L_i), or equivalently

$$\left. \begin{array}{c} \mathbf{L w} = \text{given in } \mathcal{H}, \\ \mathbf{L} \in \mathcal{L}(\mathcal{H}; \mathcal{H}), \\ (\mathbf{L w})_i = w_i + \alpha_i L_i^* \left(\rho_i^2 \sum_{j=1}^{N} L_j w_j \right). \end{array} \right\} \tag{37}$$

Then we have

Proposition 4.1 *Assume that*

$$\alpha_i = \alpha, \text{ for all } i, \tag{38}$$

and that

$$\alpha \|\rho_i - \rho_j\|_{L^\infty(\Omega)} \; \|\rho_i\|_{L^\infty(\Omega)} \text{ is small enough, for any } i, j = 1, \ldots, N. \tag{39}$$

Then $\mathbf{L}$ is invertible. In particular there is a unique Nash equilibrium of (14).

Remark 4.1 Of course, if $N = 1$ the situation is much simpler. In that case,

$$(\mathbf{L}w, w) = \|w_1\|^2 + \alpha_1 \|\rho_1 L_1 w_1\|^2,$$

hence $\mathbf{L}$ is *coercive* and so the existence and uniqueness of a *minimum* w of $J_1(v; w)$, when v is fixed, is a classical result.

Proof of Proposition 4.1: In the general case $N > 1$, one has

$$(\mathbf{Lw}, \mathbf{w}) = \sum_i \|w_i\|_{\mathcal{H}_i}^2 + \sum_i \alpha_i \left(\rho_i \sum_j L_j w_j, \rho_i L_i w_i \right). \tag{40}$$

Then one can write

$$(\mathbf{Lw}, \mathbf{w}) = \sum_{i=1}^N \|w_i\|_{\mathcal{H}_i}^2 + \alpha \left\| \sum_{i=1}^N \rho_i L_i w_i \right\|^2 + \alpha \sum_{i,j=1}^N (\rho_i - \rho_j)^2 (L_j w_j, \rho_i L_i w_i). \tag{41}$$

Applying Young's inequality, it follows that, under hypothesis (39), $\mathbf{L}$ is coercive, i.e.

$$(\mathbf{Lw}, \mathbf{w}) \geq \gamma \|\mathbf{w}\|_{\mathcal{H}}^2, \text{ for some } \gamma > 0. \tag{42}$$

The conclusion is now a consequence of the Lax–Milgram theorem.

Remark 4.2 The hypothesis (39) is certainly satisfied if $\rho_i = \rho$ for all i, in which case there is *only* one function $J_i = J_1$ for all i, and we are back to Remark 4.1 (with $\mathbf{w} = \{w_1, \ldots, w_N\}$).

4.2   Some non-existence and non-uniqueness results

We begin this subsection by some general considerations on the existence, or non-existence, of Nash equilibrium solutions.

Let $\mathcal{H}_i$, $\mathcal{K}_j$ be two families of N real Hilbert spaces $(i, j = 1, \ldots, N)$, the scalar product (or norm) in a space $\mathcal{H}$ being denoted by $(\ ,\ )_{\mathcal{H}}$ (or $\|\ \|_{\mathcal{H}}$).

We consider linear continuous operators $a_{i,j}$

$$a_{i,j} \in \mathcal{L}(\mathcal{H}_j, \mathcal{K}_i), \ \forall i, j, \tag{43}$$

and we assume that

$$a_{i,j} \text{ is compact, } \forall i, j. \tag{44}$$

We define $\mathbf{w} = \{w_1, \ldots, w_N\}$, $\mathbf{w} \in \mathcal{H} = \prod_{i=1}^N \mathcal{H}_i = \prod_{i=1}^N \mathcal{K}_i$,

$$J_i(\mathbf{w}) = \frac{1}{2} \|w_i\|_{\mathcal{H}_i}^2 \, dx dt + \frac{\alpha_i}{2} \left\| \sum_{j=1}^N a_{i,j} w_j - \eta_i \right\|_{\mathcal{K}_i}^2 \tag{45}$$

where α_i is a positive given constant, and where

$$\boldsymbol{\eta} = \{\eta_1, \ldots, \eta_N\} \text{ is given in } \prod_{i=1}^N \mathcal{K}_i. \tag{46}$$

We are looking for the Nash equilibrium points of the functionals $J_1, \ldots, J_N$. We are going to show that "in general" with respect to $\alpha = \{\alpha_i\} \in \mathbb{R}_+^N$, there *exists a unique Nash equilibrium for the functionals J_i. When α is "exceptional"* *in $\mathbb{R}_+^N$, then "in general" with respect to $\eta = \{\eta_i\} \in \prod_{i=1}^N \mathcal{K}_i$, there is no solu-* *tion. When α and η are "exceptional", there is a finite dimensional subspace of* *solutions in $\prod_{i=1}^N \mathcal{K}_i$.*

Of course, this "result" has to be made precise. An element $\mathbf{w} = \{w_1, \ldots, w_N\}$ is a Nash equilibrium iff

$$(w_i, \widehat{w_i})_{\mathcal{H}_i} + \alpha_i \left(\sum_j a_{ij} w_j - \eta_i, a_{ii} \widehat{w_i} \right)_{\mathcal{K}_i} = 0, \ i = 1, \ldots, N, \ \forall \widehat{w_i} \in \mathcal{K}_i$$

i.e.

$$a_{ii}^* \sum_{j=1}^N a_{ij} w_j + \frac{1}{\alpha_i} w_i = a_{ii}^* \eta_i, \ i = 1, \ldots, N, \tag{47}$$

where $a_{ij}^* \in \mathcal{L}(\mathcal{K}_i, \mathcal{H}_j)$ denotes the adjoint of a_{ij}.

Let us define

$$\begin{aligned} &\mathcal{A} \in \mathcal{L}\left(\prod_{i=1}^N \mathcal{H}_i, \prod_{i=1}^N \mathcal{H}_i \right), \\ &\mathcal{A}\mathbf{w} = \{a_{ii}^* \sum_{j=1}^N a_{ij} w_j\}, \end{aligned} \right\} \tag{48}$$

$$\left(\frac{1}{\alpha} \right) = \text{diagonal operator } \{w_i\} \mapsto \left\{ \frac{1}{\alpha_i} w_i \right\}, \tag{49}$$

$$\zeta_i = a_{ii} \eta_i, \ \boldsymbol{\zeta} = \{\zeta_i\}. \tag{50}$$

Then (47) is equivalent to

$$\mathcal{A}\mathbf{w} + \left(\frac{1}{\alpha} \right) \mathbf{w} = \boldsymbol{\zeta}, \ \text{in } \mathcal{H} = \prod_{i=1}^N \mathcal{H}_i, \tag{51}$$

where, by virtue of (44), $\mathcal{A}$ is compact in $\mathcal{L}(\mathcal{H}, \mathcal{H})$. Then the "result" stated above is a trivial consequence of the classical Fredholm alternative. Indeed, let us consider the $\boldsymbol{\alpha}$'s such that

$$\frac{1}{\alpha_i} = \gamma_i \lambda, \ \gamma_i \text{ fixed}, \tag{52}$$

all these numbers being positive. Then, according to the Fredholm alternative, (51) and (52) admits a *unique solution except for a countable set of λ's*. This makes precise the fact that there is, "in general" with respect to $\boldsymbol{\alpha}$, a unique solution. If λ belongs to the spectrum of $\mathcal{A} + \gamma\lambda$, then there is a solution iff $\boldsymbol{\zeta}$ is orthogonal to the null space of $\mathcal{A}^* + \gamma$, a conclusion which is "in general" not satisfied by $\boldsymbol{\zeta}$, i.e. by $\boldsymbol{\eta} = \{\eta_i\}$. If it is satisfied, then there is a finite dimensional space of solutions.

Remark 4.3 Of course, the formula (51) does *not* use the hypothesis (44). *Therefore, one has that without the hypotesis* (44) *there exists a unique Nash equilibrium if*

$$\|\alpha \mathcal{A}\|_{\mathcal{L}(\mathcal{H}, \mathcal{H})} < 1 \qquad (53)$$

(where $(\alpha \mathcal{A})\mathbf{w} = \left\{ \alpha_i a_{ii}^* \sum_j a_{ij} w_j \right\}$).

All the above remarks apply to (32), (33) if we take

$$a_{ij} = \rho_i L_j, \quad \eta_i = \rho_i \eta^T, \quad \mathcal{K}_i = L^2(\Omega), \ \forall i \qquad (54)$$

(then (53) amounts to $\alpha \left\| \rho_i - \rho_j \right\|_{L^\infty(\Omega)} \left\| \rho_i \right\|_{L^\infty(\Omega)}$ being small enough) if one verifies that L_j, as defined by

$$L_i w_i = y_i(T), \ y_i \text{ solution of (20) (with } \widehat{w_i} \text{ replaced by } w_i), \qquad (55)$$

is compact from $L^2(\mathcal{O}_i \times (0, T)) = \mathcal{H}_i$ into $L^2(\Omega)$.

If the coefficients of the operator A are smooth enough, then the solution y_i of (20) satisfies

$$y_i \in L^2(0, T : H^2(\Omega) \cap H_0^1(\Omega)), \ \frac{\partial y_i}{\partial t} \in L^2(0, T : L^2(\Omega))$$

(recall that $y_i(0) = 0$), so that $L_i \in \mathcal{L}(\mathcal{H}_i; H_0^1(\Omega))$, hence L_i is *compact* from $\mathcal{H}_i$ into $L^2(\Omega)$ (since the injection $H_0^1(\Omega) \hookrightarrow L^2(\Omega)$ is compact when Ω is bounded).

References

1. Brezis, H., 1973, *Opérateurs maximaux monotones et semigroupes de contractions dans les espaces de Hilbert*, North-Holland, Amsterdam.
2. Díaz, J. I. and Lions, J. L., 1998, article in preparation.
3. Gabay, D. and Lions, J. L., 1994, Décisions stratégiques à moindres regrets, *C. R. Acad. Sci. Paris*, t. **319**, Série I, 1049–1056.
4. Gilbarg, D. and Trudinger, N. S., 1977, *Elliptic Partial Differential Equations of Second Order*, Springer, Berlin.
5. Mizohata, S., 1958, Unicité du prolongement des solutions pour quelques opérateurs différentiels paraboliques, *Mem. Coll. Sci. Univ. Kyoto*, Ser. A31, **3**, 219–239.
6. Lebeau, G. and Robbiano, L., 1995, Contrôle exact de l'équation de la chaleur, *Communications in PDE*, **20**, 335–356.
7. Lions, J. L., 1981, *Some Methods in the Mathematical Analysis of Systems and Their Control*, Science Press and Gordon and Breach.
8. Lions, J. L., 1994, Some Remarks on Stackelberg's Optimization, Mathematical Models and Methods in Applied Sciences, **4**, no. 4, 477–487.
9. Lions, J. L., 1986, Contrôle de Pareto de systèmes distribués: Le cas d'évolution, *C. R. Acad. Sci. Paris*, t. **302**, Série I, 413–417.
10. Saut, J. C. and Scheurer, B., 1987, Unique Continuation for Some Evolution Equations, *J. Differential Equations*, **66**, 118–139.
11. Simon, J., 1987, Compact Sets in the Space $L^p(0, T; B)$, *Annali di Matematica Pura ed Applicata* (IV), **CXLVI**, 65–96.
12. Stackelberg, H. von, 1934, *Marktform und Gleichgewicht*, Springer, Berlin.

3D Simulation in the lower troposphere: wind field adjustment to observational data and dispersion of air pollutants from combustion of sulfur-containing fuel

G. Winter, J. Betancor, and G. Montero

Escuela Técnica Superior de Ingenieros Industriales, Universidad de Las Palmas,
Edificio de Ingenierías, Campus Universitario de Tafira Baja,
35017 Las Palmas de Gran Canaria, Spain

1 Introduction

Combinations of mathematical models with data at particular points from observational networks are required in order to generate physically consistent wind fields and atmospheric pollutant distributions. We describe a methodology used to evaluate the modifications to wind flow and pollutant dispersion, mainly caused by the interaction of the air flow with the terrain.

The troposhere extends from the ground until an average altitude of 11 Km. We focus our attention on the part of the troposphere that is directly influenced by the presence of the earth's surface, a region where surface-atmosphere turbulent exchange processes take place, the so-called planetary boundary layer (PBL). Often this layer corresponds to the lowest 500–1500 m of the atmosphere, which is the most important region from many viewpoints, as pollutant emission, frictional drag or terrain induced flow modification. Indirectly, the whole troposphere can change in response to surface characteristics, but this response is relatively slow outside of the boundary layer. The thickness of this region is quite variable in space and time due to the thermal stability conditions, ranging between 100 m at night time with light wind and turbulence to 1–2 Km on sunny days with surface heating. It is usually assumed that the boundary layer includes a statement about one-hour or less timescales. On the other hand, if the ground surface is not spatially homogeneous (as it is the case in general), this inhomogeneity is reflected in the PBL. For this reason, the context of numerical simulation in 3D is of interest. Mean wind is responsible for very rapid horizontal transport or advection. Wind velocity increases from 0 to about 70% of its maximum PBL value, while the wind direction is nearly constant with height.

Within the PBL, in particular near the ground surface, typically up to 10–100 m, we have the surface layer (SF). It is usually assumed that the SL covers the bottom 10% of the PBL. In this layer, where the wind is influenced by the prevailing high-level flows and the effect of the surface is well felt, the wind is mainly determined by the nature of the surface and the vertical temperature gradient. Effects of density stratification are small and the wind speed follows a nearly logarithmic vertical profile, even under stable and unstable condition forms of the velocity profiles. These profiles are very useful because of the stratified-boundary-layer conservation equations, which are difficult to solve due to the closure problem in the turbulence models.

Within the SF layer, in the vicinity of the ground surface, turbulence is strongly affected by roughness and viscous effects may become significative. Immediately adjacent to the surface, a laminar sublayer (also called interfacial layer, or microlayer) is identified, in which strong molecular viscosities become important. However, the thickness of this layer is typically less than a centimeter and for all purposes it can be ignored. Above the SL, the role of the Coriolis force becomes relevant with respect to the friction forces. The wind velocity changes slowly while the wind direction veers describing the so-called Ekman spiral. This part of the PBL is called Ekman layer. In these layers, some important simplifications in the equations of continuity, motion and energy can be made, as the continuity equation for an incompressible fluid.

At the top of the PBL, the flow is nearly independent of the nature of the surface, and above of this top the region is called free atmosphere. This layer is usually called the geostrophic layer, where motion of air approximates that of an inviscid fluid in laminar flow and the direction of winds is mainly determined by horizontal pressure gradients and Coriolis forces.

An accurate estimation of wind and atmospheric pollutant distributions requires to use mathematical models linked with meteorological observations. We treat a model of wind field modelisation on complex terrain, e.g. adjusting meteorology and topography data with small computational effort in 3D. This adjustment model is a so-called mass-consistent model (MMC), which satisfies the following: the vorticity of the observed wind field is conserved by the adjustment rotational; the flow velocity field is nondivergent (under incompressible conditions corresponds to the continuity equation); and an impermeability condition holds at the ground. This velocity field adjustment model is characterized by a mixed variational formulation as a result of the corresponding optimization problem, defined and solved by looking for a saddle point of the associated Lagrangian function. This model fits the available experimental measurements, and their mixed variational formulation is very suitable, since the numerical solution exactly satisfies divergence-free conditions pointwise.

The purpose is to provide a realistic methodology for simulation of winds, where the boundary conditions and initial velocity field are constructed in a consistent way from experimental data with the use of different sources, some ground stations, geostrophic wind, atmospheric stability class, roughness parameter and the dependence of wind speed on height given by appropriate logarithmic profiles for each of the above-mentioned atmospheric layers. Numerical results are showed with the MMC model in a region of Canary Islands with real data.

The 3D Navier–Stokes (NS) equation needs boundary conditions, and an initial velocity field has to be specified. We highlight that a possible alternative methodology to provide the initial velocity field and boundary conditions to the

3D NS formulation can be established from results obtained with the wind field adjustment model.

In what follows we consider some aspects on two numerical methods to solve coupled convection-difussion equations for modelling of air pollutants: one of them based on characteristic lines and another one on a Taylor–Galerkin procedure. Some results of numerical stability and consistence are compared. We especially focus on modelling oxidation and hydrolysis of sulfur and nitrogen oxides released to the surface layer, which, once oxidated, are major contributors to acid rain in geographical regions, and producing aerosols with proved climatic implications. Nitrogen oxides play an important role in the atmospheric photochemistry of other greenhouse gases. The dry deposition process is represented by the so-called deposition velocity, which is proportional to the degree of absorptivity of the surface, and it is assumed to be a proportional constant between vertical flow and concentration, and thus it is treated as a boundary condition. The wet deposition is considered as a source term in the convection-diffusion equation using the washout coefficient. A numerical application considering the same topography and wind field and relative to calculate the distribution of concentrations of sulfure oxide and sulfate is presented.

2   Wind field adjustment model

For a given bounded open three-dimensional domain with boundary $\Gamma = \Gamma_1 \cup \Gamma_2$, we look for a field $\boldsymbol{u}$ that adjusts, in a least square sense, to a velocity field $\boldsymbol{u}_0$, obtained from the interpolation of experimental measurements and vertical extrapolation by suitable profiles for each atmospheric layer within the PBL, and verifying

$$\boldsymbol{\nabla} \cdot \boldsymbol{u} = 0 \ \text{ on } \Omega$$
$$\boldsymbol{u} \cdot \boldsymbol{n} = 0 \ \text{ at } \Gamma_1, \tag{1}$$

where $\Gamma_1 = \Gamma_t \cup \Gamma_u$, with Γ_u the upper altitude of the PBL, and Γ_t the surface of terrain. A zero flux boundary condition is used for the remaining boundaries Γ_2. The least square functional to be minimized is:

$$J(\boldsymbol{u}) = \frac{1}{2} \int_{\Omega} (\boldsymbol{u} - \boldsymbol{u}_0)^t \cdot \boldsymbol{P} \cdot (\boldsymbol{u} - \boldsymbol{u}_0) \, d\Omega + \frac{\beta}{2} \int_{\Gamma_2} \boldsymbol{n} \cdot (\boldsymbol{u} - \boldsymbol{u}_0)^2 \, d\Gamma, \tag{2}$$

where $\boldsymbol{P}$ denotes a diagonal matrix. Different values of their entries (the Gauss precision moduli) allow ponderation between horizontal and vertical velocity components (usually less value with relation to the vertical component). Then, the wind field will be a solution of the following problem:

"Find $\boldsymbol{u} \in K$ that verifies

$$J(\boldsymbol{u}) = \min_{\boldsymbol{v} \in K} J(\boldsymbol{v})$$

with $K = \{\boldsymbol{v}; \ \boldsymbol{\nabla} \cdot \boldsymbol{v} = 0, \ \boldsymbol{v} \cdot \boldsymbol{n}|_{\Gamma_1} = 0\}$."

The problem can be formulated as a saddle point problem for the Lagrangian:

$$L(\boldsymbol{v}, q) = J(\boldsymbol{v}) + \int_\Omega q \cdot \boldsymbol{\nabla} \cdot \boldsymbol{v} \, d\Omega.$$

More precisely, if $L^2(\Omega)$ is the space of square integrable functions and $H^1(\Omega)$ the subspace of $L^2(\Omega)$ with square integrable first derivatives, we denote:

$$H^1_{0,\Gamma_2}(\Omega) = \left\{ \varphi \in H^1(\Omega); \, \varphi|_{\Gamma_2} = 0 \right\},$$
$$H(\boldsymbol{\nabla}, \Omega) = \left\{ \boldsymbol{v} \in \left(L^2(\Omega)\right)^d; \, \boldsymbol{\nabla} \cdot \boldsymbol{v} \in L^2(\Omega) \right\},$$

and, by introducing the space of vector functions such that $\boldsymbol{v} \cdot \boldsymbol{n} = 0$ on Γ_1 in a meaningful way, say,

$$H_{0,\Gamma_1}(\boldsymbol{\nabla}, \Omega) = \left\{ \boldsymbol{v} \in H(\boldsymbol{\nabla}, v); \, \int_\Gamma \varphi \boldsymbol{v} \cdot \boldsymbol{n} \, d\Gamma = 0 \, \forall \varphi \in H^1_{0,\Gamma_2}(\Omega) \right\}, \quad (3)$$

we search for the couple $(\boldsymbol{u}, \lambda) \in H_{0,\Gamma_1}(\boldsymbol{\nabla}, \Omega) \times L^2(\Omega)$ such that

$$L(\boldsymbol{u}, q) \leq L(\boldsymbol{u}, \lambda) \leq L(\boldsymbol{v}, \lambda)$$

for all $q \in L^2(\Omega)$ and all $\boldsymbol{v} \in H_{0,\Gamma_1}(\boldsymbol{\nabla}, \Omega)$, which is characterized by

$$\frac{\partial L(\boldsymbol{u}, \lambda)}{\partial v} = 0 \text{ and } \frac{\partial L(\boldsymbol{v}, \lambda)}{\partial q} = 0 \text{ for all } \boldsymbol{v} \in H_{0,\Gamma_1}(\boldsymbol{\nabla}, \Omega), \quad (4)$$

$$\int_\Omega q \cdot \boldsymbol{\nabla} \cdot \boldsymbol{u} \, d\Omega = 0 \text{ for all } q \in L^2(\Omega), \quad (5)$$

obtaining

$$\int_\Omega \boldsymbol{v}^t \cdot \boldsymbol{P} \cdot (\boldsymbol{u} - \boldsymbol{u}_0) \, d\Omega + \int_\Omega \lambda \boldsymbol{\nabla} \cdot \boldsymbol{v} \, d\Omega + \beta \int_{\Gamma_2} \boldsymbol{v} \cdot \boldsymbol{n} \, (\boldsymbol{n} \cdot (\boldsymbol{u} - \boldsymbol{u}_0)) \, d\Gamma = 0. \quad (6)$$

The variational formulation given by (5) and (6) can be solved with mixed finite elements (see [6] for more details), with great advantages. However, one more classical formulation, known as matrix mass-consistent model (MMC), is used instead. It can be derived from (6) with the assumption that the Lagrange multiplier be sufficiently regular (see [2]), and then we obtain

$$\boldsymbol{u} = \boldsymbol{u}_0 + \boldsymbol{P}^{-1} \boldsymbol{\nabla} \lambda. \quad (7)$$

Now, the problem to solve is

$$\boldsymbol{\nabla} \cdot (\boldsymbol{u} - \boldsymbol{u_0}) = \boldsymbol{\nabla} \cdot \left(\boldsymbol{P}^{-1} \boldsymbol{\nabla} \lambda\right), \quad (8)$$

subject to the following boundary conditions:

$$-\boldsymbol{P}^{-1} \cdot \frac{\partial \lambda}{\partial n} = \boldsymbol{n} \cdot \boldsymbol{u}_0 \text{ on } \Gamma_1$$

$$-\boldsymbol{P}^{-1} \cdot \frac{\partial \lambda}{\partial n} = \frac{\lambda}{\beta} \quad \text{on } \Gamma_2. \tag{9}$$

The standard finite element method can be used to obtain the Lagrange multiplier from (8) subject to the boundary conditions (9), and then the velocity field $\boldsymbol{u}$ from (7). This traditional procedure is usually considered, but the field thus obtained is discontinuous through the faces of the finite elements and does not satisfy the incompressibility condition pointwise.

2.1   Initial wind field

The construction of the initial wind field is the most important and critical step in mass-consistent models, since it introduces the experimental data into the model at each node of the computational mesh. Usually, observational data are available at 10 m above the terrain at different locations, from sensors. We propose a two-step procedure with the interpolation of a velocity field at x_{3_e} over terrain as a first step using the following expression (see [5] and [8]):

$$\boldsymbol{u}_0(x_{3_e}) = \varepsilon \frac{\sum_{i=1}^{n} \frac{\boldsymbol{u}_i}{d_i^m}}{\sum_{i=1}^{n} \frac{1}{d_i^m}} + (1 - \varepsilon) \frac{\sum_{i=1}^{n} \frac{\boldsymbol{u}_i}{|\Delta h_i|}}{\sum_{i=1}^{n} \frac{1}{|\Delta h_i|}}, \tag{10}$$

where d_i represents the horizontal distance from the ith station to the point considered, $|\Delta h_i|$ corresponds to the height differences between them, and n is the number of observation stations. The use of the parameter ε, such that $0 \leq \varepsilon \leq 1$, will allow us to balance the contribution of both weights of interpolation: horizontal distance and height differences. In practical applications, good results were obtained with $m = 2$.

Once the velocity field has been interpolated at x_{3_e} over terrain, it is vertically extrapolated using different profiles in every layer where PBL is considered to be subdivided by similarity theory, taking into account stability data obtained from vertical soundings. Thus, within the surface layer the velocity is computed at different heights by (see [14]):

$$\boldsymbol{u}_0(x_3) = \frac{\boldsymbol{u}_*}{\kappa} \cdot \begin{cases} \ln \dfrac{x_3}{x_{3_0}} + \dfrac{4.7}{L}(x_3 - x_{3_0}) & \zeta > 0 \text{ stable} \\[2ex] \ln \dfrac{x_3}{x_{3_0}} & \zeta = 0 \text{ neutral} \\[2ex] \ln \dfrac{x_3}{x_{3_0}} + 2\left(\dfrac{1}{\tan \eta_r} - \dfrac{1}{\tan \eta_0}\right) + \\[2ex] +\ln \dfrac{(\eta_0^2 + 1)(\eta_0 + 1)^2}{(\eta_r^2 + 1)(\eta_r + 1)^2} & \zeta < 0 \text{ unstable,} \end{cases} \tag{11}$$

with

$$\eta_{\mathrm{r}} = \left(1 - 15\frac{x_3}{L}\right)^{1/4} \quad \eta_0 = \left(1 - 15\frac{x_{3_0}}{L}\right)^{1/4} \tag{12}$$

where κ is the Von Karman parameter (usually equal to 0.4), L the Monin–Obukhov length, and x_{3_0} the characteristic length of rugosity, which represents the depth of the laminar sublayer adjacent to the earth's surface. Its value will depend on the surface itself.

In the Ekman layer, the velocity vector is computed using a linear interpolation formula between the wind field at the top of the Surface Layer, computed by (11), and the geostrophic wind, as suggested by Troen (see [5]). A third-order weight function is used:

$$\rho(x_3) = 1 - \left(\frac{x_3 - x_{3_{\mathrm{sl}}}}{x_{3_{\mathrm{PBL}}} - x_{3_{\mathrm{sl}}}}\right)^2 \cdot \left(3 - 2\frac{x_3 - x_{3_{\mathrm{sl}}}}{x_{3_{\mathrm{PBL}}} - x_{3_{\mathrm{sl}}}}\right).$$

Over the PBL, in the geostrophic area, the velocity vector is consider to be constant with height u_{g}, computed from soundings data.

The atmospheric stability will be characterized by the Monin–Obukhov length L, which is related to the Richardson flux dimensionless number Rf. Thus, positive values for L denote a stable atmosphere while negative values represent an unstable atmosphere. For $L \to \infty$, the atmosphere is said to be in neutral condition. Several methods have been developed to compute L from directly observable data. We use the Golder method (see [3]) to compute L from

$$\frac{1}{L} = a + b \log x_{3_0}, \tag{13}$$

where a and b are two coefficients whose values are indicated in Table 1 as a function of the corresponding Pasquill stability classes (see [4]), which can be determined from observational data of incoming solar radiation, mean wind speed, and night-time cloud-cover fraction.

Table 1. Coefficients for (13) as a function of Pasquill stability classes (see [10]).

Atmospheric condition	Pasquill stability class	Coefficients	
		a	b
Extremely unstable	A	−0.096	0.029
Moderately unstable	B	−0.037	0.029
Slightly unstable	C	−0.002	0.018
Neutral	D	0	0
Slightly stable	E	0.004	−0.018
Moderately stable	F	0.035	

As scale parameter, the modulus of the friction velocity vector $|\boldsymbol{u}_*|$ is used. Thus the height of the PBL can be evaluated by the expression

$$x_{3_{PBL}} = \frac{\gamma\,|\boldsymbol{u}_*|}{f},$$

where f is the Coriolis parameter, γ is a constant and its value depends on atmospheric stability, usually between 0.15 and 0.35.

The maximum height of the mixed layer can be evaluated by

$$h = \begin{cases} \gamma'\sqrt{\dfrac{u_*}{f}L} & \text{stable} \\[2ex] x_{3_{PBL}} & \text{non stable}, \end{cases}$$

where γ' is a parameter whose value is 0.4. The height of the surface layer is here assumed to be

$$x_{3_{sl}} = \frac{h}{10}.$$

2.2   Parameters of ponderation

Atmospheric stability is also considered in order to evaluate the Gauss precision moduli α_i, $i = 1, 2, 3$. Identical Gauss precision moduli are generally considered for the horizontal directions and a unique parameter is then considered:

$$\alpha^2 = \frac{\alpha_v^2}{\alpha_h^2} = \frac{\tau_v}{\tau_h} = \tau,$$

which is called transmissivity coefficient. Many authors have proposed different methods for the calculation of the optimal value of this parameter, which is, in fact, very case-dependent. For our model, we have choosen the method proposed by Moussiopoulos [13], which considers the Strouhal dimensionless number Str, as the most representative parameter which takes into account the effects of the atmospheric stability through the buoyancy frequency $\mathcal{N}$, and the orography through the characteristic height difference $\mathcal{H}$. The Strouhal dimensionless number is defined by

$$\text{Str} = \begin{cases} \dfrac{\mathcal{N}\mathcal{H}}{u_*} & \text{stable and neutral} \\[3ex] -\dfrac{\mathcal{N}\mathcal{H}}{u_*} & \text{unstable}, \end{cases}$$

where

$$\mathcal{N} = \sqrt{\frac{g}{\theta}\left|\frac{d\theta}{dz}\right|} \qquad \text{and} \qquad \mathcal{H}_i = \frac{\displaystyle\sum_{j\neq i}\frac{|\Delta h_{ij}|}{d_{ij}^2}}{\displaystyle\sum_{j\neq i}\frac{1}{d_{ij}^2}}.$$

Thus, α must be computed in stable and neutral conditions by

$$\alpha^2 = 1 - \frac{\text{Str}^4}{2}\left(\sqrt{1 + 4\,\text{Str}^{-4}} - 1\right). \tag{14}$$

For unstable conditions, the following relationship must be satisfied. However, it is observed that the parametrization for unstable stratification has no significant influence on the resulting wind field.

$$1 < \alpha^2 < \left[\alpha^2(-\text{Str})\right]^{-1}. \tag{15}$$

3   Pollution transport model

We consider the atmospheric convection-diffusion equation, which provides a more appropriated model than the Gaussian models, because of the ability to include changes in wind speed and variable eddy diffusivities. Dealing with gaseous pollutants, it is assumed that their concentration does not affect the meteorology at some extent, and the equation of continuity can be solved independently of the coupled momentum and energy equations. Under appropriate restrictions, the flux of pollutants is proportional to the gradient of its mean concentration. Thus, in a three-dimensional, inhomogeneous environment, we will have the simplified equation:

$$\frac{\partial c_i}{\partial t} + \boldsymbol{\nabla} \cdot (\boldsymbol{u} \cdot c_i) - \boldsymbol{\nabla} \cdot (\boldsymbol{K}_i \cdot \boldsymbol{\nabla} \cdot c_i) = f_i \tag{16}$$

where $c_i = c_i(\boldsymbol{x}, t)$ is the mean concentration of the ith atmospheric pollutant species and $\boldsymbol{K}_i$ is a diagonal matrix with difussion coefficients. The set of equations (16) (one for each considered pollutant) are to be solved subject to the following initial and boundary conditions:

$$
\begin{aligned}
c_i &= c_{i_0}(\boldsymbol{x}, 0) & &\text{at } t = 0 \\
c_i &= c_{i_2}(\boldsymbol{x}) & &\text{on } \Gamma_2 \\
-\boldsymbol{K}_i(\boldsymbol{x}) \cdot \boldsymbol{\nabla} c_i \cdot \boldsymbol{n} &= v_{d_i}(\boldsymbol{x}) \cdot c_i & &\text{on } \Gamma_t \\
\boldsymbol{\nabla} c_i \cdot \boldsymbol{n} &= 0 & &\text{on } \Gamma_u,
\end{aligned}
\tag{17}
$$

where $c_{i_0}(\boldsymbol{x}, 0)$ is the initial concentration field of the ith pollutant, $c_{i_2}(\boldsymbol{x})$ its concentration along the frontier Γ_2, and v_{d_i} the deposition velocity parameter, which models the vertical flux downward of the ith species above the surface, known as *dry deposition*.

3.1   Dry deposition

It refers to the vertical flux downward the earth's surface of both gaseous and particulate pollutants, which are absorbed by soil, water or vegetation. Its complex mechanism can be viewed as consisting of three consecutive steps, each with its own resistance to the flow: Transport from the surface layer to the vicinity of the earth's surface, diffusion through laminar sublayer, and absorption or transfer to surface recipients. This flux can be modelled by the concentration of the pollutant at some height just above the surface multiplied by the deposition velocity v_{d_i}, which is a function of the species to be removed, meteorological properties on the surface layer, and the degree of absorptivity of the surface itself. Functionally, this flux can be expressed as

$$\mathcal{D}_i = v_{d_i} c_i(x_1, x_2, x_{3_{\text{earth}}}, t)$$

and, thus, will be included as a boundary condition at the earth's surface.

3.2   Source term

We assume that it is composed of three terms for every species i:

$$f_i = E_i + R_i + P_i,$$

where $E_i(\boldsymbol{x}, t)$ includes continuous and instantaneous emission of the ith species to the atmosphere, $R_i(c_1, \ldots, c_n, t)$ represents the rate of generation or elimination of the ith species through chemical reactions, and $P_i(\boldsymbol{x}, t)$ accounts for the rate of elimination of the ith species through wet deposition because of water absorption of pollutants during rain or other cloud processes.

To model the emission rate E_i of the ith pollutant to the atmosphere from a power plant, a point source function is assumed:

$$E_i = e_{i_0}(t)\delta(x_1 - x_{1_0})\,\delta(x_2 - x_{2_0})\,\delta(x_3 - x_{3_0}), \tag{18}$$

where $(x_1, x_2, x_3)_0$ are the location coordinates of the power plant. This expression leads us to consider an instantaneous emission at $t = t_0$ with

$$e_{i_0}(t) = e_{i_0}\delta(t - t_0)$$

and, thus, by the application of the superposition principle, any combination of continous and instantaneous sources. The adopted models for the wet deposition and the chemical reactions are now discussed in detail.

3.3   Wet removal processes

Wet deposition accounts for the removal processes of pollutants that take place in the atmosphere because of their absorption by clouds, rainfall water, etc.

We assume wet deposition as a first-order process. It is assumed that the rate of removal depends linearly on the airborne concentration of the material and is independent of the quantity of material previously scavenged. Thus, the local rate of removal of the ith gas is given by

$$P_i = -\Lambda_i(x_3, t)\,c_i(x_1, x_2, x_3, t)$$

where $\Lambda_i(x_3, t)$ is the washout coefficient of the ith gas, which, in general, will depend on the height over the earth's surface. With the assumption that the material being scavenged is uniformly distributed vertically in the layer of depth h or the mixed layer height, a wet deposition velocity v_{w_i} is defined by analogy with dry deposition (see [11]) as

$$v_{w_i} = \overline{\Lambda} h,$$

which can be expressed in terms of more usual parameters such as the washout ratio w_{r_i}, and the precipitation intensity p_0, whose value is affected by the dominant atmospheric condition ranging from 0.5 to 25 mm · h^{-1} (see [10]).

3.4   Chemical transformations

We focus our attention on the oxides of sulfur and nitrogen which are typically released to the atmosphere in the combustion process of any industrial fuel. These species, once released, are oxidized to their corresponding acids, which are largely responsible for the acid rain. The set of reactions that involve nitrogen and sulfur oxides are summarized in Fig. 1. It shows the coupling between mechanisms of reaction that involve families of both sulfur and nitrogen compounds, through shared radicals OH· and HO$_2$·. Most of these reactions are photochemically excited and involve a great number of intermediate species of short lifetime, such as SO$_2$· or HOSO$_2$·.

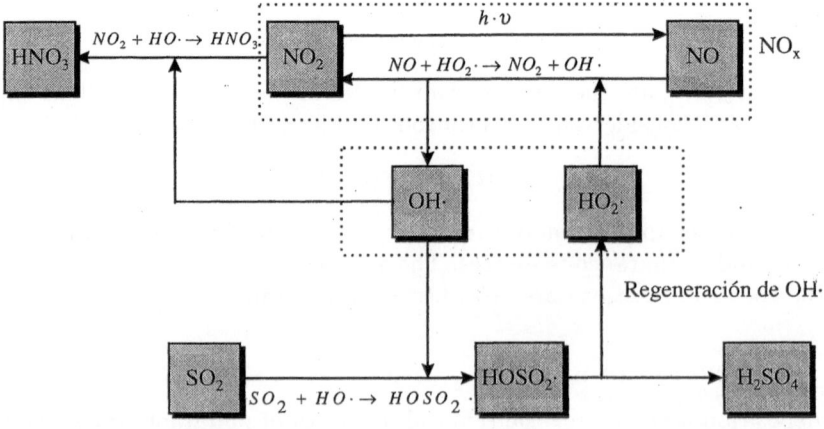

Fig. 1. Set of chemical reactions involving OH· and HO$_2$·.

Two relevant mechanisms are remarked in the dotted boxes of Fig. 1: One of them represents the chemical equilibrium of the oxides of nitrogen, and the other one leads to the regeneration of radicals OH· and HO$_2$·, which will allow us to uncouple the mechanisms of reaction of both families of pollutants. Certainly, the oxidation of SO$_2$ by HO$_2$· is so slow when compared with the reactions that involve the chemical equilibrium of nitrogen oxides that it is assumed as first approximation that its concentration remains nearly constant because of these reactions. With these assumptions, the rates of reaction of the final species are:

$$\frac{d[\text{NO}_x]}{dt} = -2k_1k_8[\text{NO}_x] \tag{19}$$

$$\frac{d[\text{HNO}_3]}{dt} = 2k_1k_8[\text{NO}_x] \tag{20}$$

$$\frac{d[\text{SO}_2]}{dt} = -2\frac{k_1k_9}{k_8}[\text{SO}_2] \tag{21}$$

$$\frac{d[\text{H}_2\text{SO}_4]}{dt} = 2\frac{k_1k_9}{k_8}[\text{SO}_2] \tag{22}$$

which are linear respect to the concentration of the pollutants from its corresponding family. In a more general representation,

$$R_i = \sum_{j=1}^{N} \alpha_{ij}c_j$$

with $i, j = 1, 2, \ldots, N$, where α_{ij} are the specific rates of generation or elimination of the ith species in the different reaction mechanisms.

In the following, we will collect the terms related to the chemical transformations and wet deposition because of their functional similarity in a term

$$R_i' = R_i + P_i = \sum_{j=1}^{N} \alpha_{ij}'c_j,$$

so the source term will be referred as $f_i = E_i + R_i'$.

3.5   Numerical solution method

About the numerical method to solve the problem (16), we describe two procedures: One related with the characteristic lines of flow, and the another one is a Taylor–Galerkin scheme. We summarize the establishment of both numerical schemes particularized to our air pollution problem. We have implemented both schemes with a finite element method in 3D to carry out the applications.

Characteristic lines scheme: From the transient equation

$$\frac{dc_i}{dt} - \boldsymbol{\nabla} \cdot (\boldsymbol{K}_i \cdot \boldsymbol{\nabla} \cdot c_i) = f_i \tag{23}$$

we can establish a discrete treatment of the material derivative in the implicit scheme

$$c_i^{n+1}(\boldsymbol{x}) - \Delta t\boldsymbol{\nabla} \cdot (\boldsymbol{K}_i \cdot \boldsymbol{\nabla} \cdot c_i^{n+1}) = \Delta t f_i^{n+1} + c_i^n(\hat{\boldsymbol{x}}),$$

where $\hat{\boldsymbol{x}}$ is the position of a fluid particle on the characteristic line at a time t_n. Now we can expand in a Taylor series the concentration $c_i^n(\hat{\boldsymbol{x}})$ at a location $\hat{\boldsymbol{x}}$ from the location $\boldsymbol{x}$ with concentration $c_i^n(\boldsymbol{x})$ through a characteristic line of flow, which results in

$$c_i^n(\hat{\boldsymbol{x}}) = c_i^n(\boldsymbol{x} - \Delta\boldsymbol{x}) = u^n(\boldsymbol{x}) - \sum_{p=1}^{3} \Delta x_p \frac{\partial c_i^n(\boldsymbol{x})}{\partial x_p} + \qquad (24)$$

$$+\frac{1}{2} \sum_{p=1}^{3}\sum_{q=1}^{3} \Delta x_p \Delta x_q \frac{\partial^2 c_i^n(\boldsymbol{x})}{\partial x_p \partial x_q} + O(\|\Delta\boldsymbol{x}\|^3).$$

Expanding the position vector at $\hat{\boldsymbol{x}}$ so that every coordinate p of the position vector is

$$\hat{x}_p = x_p - \Delta t \boldsymbol{u}(\boldsymbol{x}) \cdot \boldsymbol{\nabla} u_p \qquad (25)$$

and substituting (25) in (24), we can evaluate all its terms at a location $\boldsymbol{x}$ at different times by

$$c_i^{n+1} - \Delta t \boldsymbol{\nabla} \cdot \left(\boldsymbol{K}_i \cdot \boldsymbol{\nabla} \cdot c_i^{n+1}\right) - \Delta t \sum_{j=1} \alpha'_{ij} c_j^{n+1} = \Delta t E_i^{n+1} + c_i^n -$$

$$-\Delta t \boldsymbol{u} \cdot \boldsymbol{\nabla} c_i^n + \frac{\Delta t^2}{2} \sum_{p=1}^{3} \boldsymbol{u} \cdot \boldsymbol{\nabla} u_p \frac{\partial c_i^n}{\partial x_p} + \frac{\Delta t^2}{2} \sum_{p=1}^{3}\sum_{q=1}^{3} u_p u_q \frac{\partial^2 c_i^n(\boldsymbol{x})}{\partial x_p \partial x_q} + O(t^3),$$

which can be treated numerically by standard finite element techniques. A similar treatment can be given to (27) in an explicit or even semiimplicit scheme.

Taylor–Galerkin scheme: The Taylor–Galerkin scheme proposed here is a variant that introduces an additional term respect to other classical schemes of this kind. We consider the following Taylor approximation:

$$c_i^{n+1} = c^n + \Delta t \left.\frac{\partial c_i}{\partial t}\right|_n + \frac{\Delta t^2}{2} \left.\frac{\partial^2 c_i}{\partial t^2}\right|_{n+\Theta} + O(\Delta t^3), \qquad (26)$$

where Θ is a number such that $0 \leq \Theta \leq 1$. The first derivative of the pollutant concentration can be obtained from (16), while for the second derivative we evaluate

$$\left.\frac{\partial^2 c_i}{\partial t^2}\right|_{n+\Theta} = \frac{\partial}{\partial t}\left[\frac{\partial c_i}{\partial t}\right]_{n+\Theta} = -\frac{\partial}{\partial t}\left[\boldsymbol{\nabla} \cdot (\boldsymbol{u} \cdot c_i) - \boldsymbol{\nabla} \cdot (\boldsymbol{K}_i \cdot \boldsymbol{\nabla} \cdot c_i) - f_i\right]_{n+\Theta}$$
$$(27)$$

whose terms can be evaluated as follows:

$$\frac{\partial}{\partial t}\left[\boldsymbol{\nabla} \cdot (\boldsymbol{u} \cdot c_i)\right] = -\boldsymbol{\nabla} \cdot (\boldsymbol{u} \cdot (\boldsymbol{\nabla} \cdot (\boldsymbol{u} \cdot c_i) - \boldsymbol{\nabla} \cdot (\boldsymbol{K}_i \cdot \boldsymbol{\nabla} \cdot c_i) - f_i))$$

$$\frac{\partial f_i}{\partial t} = \frac{\partial E_i}{\partial t} - \sum_{j=1}^{N} \alpha'_{ij} \left(\boldsymbol{\nabla} \cdot (\boldsymbol{u} \cdot c_i) - \boldsymbol{\nabla} \cdot (\boldsymbol{K}_i \cdot \boldsymbol{\nabla} \cdot c_i) - f_i \right) \tag{28}$$

$$\frac{\partial}{\partial t} \left[\boldsymbol{\nabla} \cdot (\boldsymbol{K}_i \cdot \boldsymbol{\nabla} \cdot c_i) \right] = \boldsymbol{\nabla} \cdot \left[\boldsymbol{K}_i \cdot \boldsymbol{\nabla} \cdot \frac{\partial c_i}{\partial t} \right] = \boldsymbol{\nabla} \cdot \left[\boldsymbol{K}_i \cdot \boldsymbol{\nabla} \cdot \frac{\partial f_i}{\partial t} \right],$$

where third or higher degree derivatives have been ignored.

By substitution of (28) in (27), a final expression for the numerical scheme is obtained:

$$
\begin{aligned}
c_i^{n+1} = {} & c_i^n - \Delta t \left(\boldsymbol{\nabla} \cdot (\boldsymbol{u} \cdot c_i) - f_i \right)_n - \Delta t \left(\boldsymbol{\nabla} \cdot (\boldsymbol{K}_i \cdot \boldsymbol{\nabla} \cdot c_i) \right)_{n+\Theta} \\
& + \frac{\Delta t^2}{2} \left(1 - 2\Theta \right) \left\{ \boldsymbol{\nabla} \cdot (\boldsymbol{K}_i \cdot \boldsymbol{\nabla} \cdot f_i) \right\}_{n+\Theta} \\
& + \frac{\Delta t^2}{2} \left\{ \boldsymbol{\nabla} \cdot (\boldsymbol{u} \cdot (\boldsymbol{\nabla} \cdot (\boldsymbol{u} \cdot c_i) - f_i)) \right. \\
& \left. - \sum_{j=1}^{N} \alpha'_{ij} \left(\boldsymbol{\nabla} \cdot (\boldsymbol{u} \cdot c_i) - \boldsymbol{\nabla} \cdot (\boldsymbol{K}_i \cdot \boldsymbol{\nabla} \cdot c_i) - f_i \right) \right\}_{n+\Theta}
\end{aligned}
$$

whose terms evaluated at time $n + \Theta$ will be approximated by weighted interpolation between time n and time $n + 1$. The expression obtained can be treated directly with standard finite element techniques.

When considering explicit schemes, $\Theta = 0$, identical results will be obtained with both characteristic lines and the Taylor–Galerkin method, which means that the Taylor–Galerkin method allows us to establish a scheme to solve numerically the convection-diffusion problem without considering the nature of the fluid or media in which the process takes place, extending its validity to problems in which the concept of fluid particle is meaningless.

3.6 Stability and accuracy

In this section, we analyse the stability and accuracy of the adopted scheme for solving convection-diffusion equation. The analytical solution of the problem

$$\frac{\partial c}{\partial t} + \boldsymbol{\nabla} \cdot (\boldsymbol{u} \cdot c) - \boldsymbol{\nabla} \cdot (\boldsymbol{K} \cdot \boldsymbol{\nabla} \cdot c) = \alpha c \tag{29}$$

subject to initial and boundary conditions (17) may be expanded in Fourier series, each mode m having the expression

$$c_m (\boldsymbol{x}, t) = F_m e^{-\delta_m t + (\boldsymbol{\xi}_m \cdot \boldsymbol{x} - \Omega_m t) i},$$

where F_m is the amplitude of the mode, $\boldsymbol{\xi}_m$ the wave vector, $\Omega_m = \boldsymbol{v} \cdot \boldsymbol{\xi}_m$ the frequency of the mode m, and $\delta_m = K \left| \boldsymbol{\xi}_m \right|^2 - \alpha$ is the damping. We will use a linear finite element approximation in the regular mesh represented in Fig. 2.

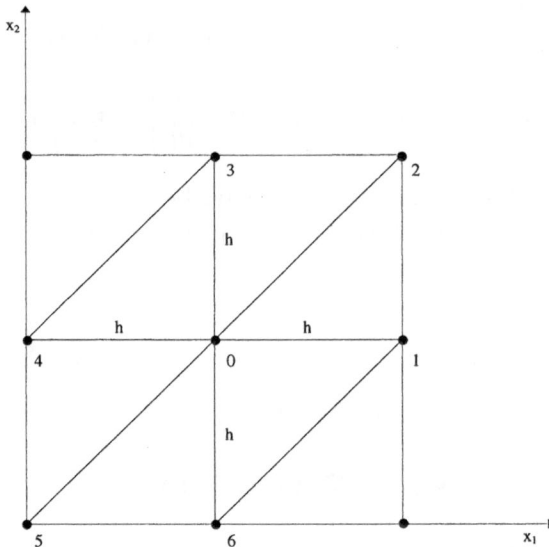

Fig. 2. Regular 2D mesh used for the study of accuracy and stability.

Characteristic lines scheme: Considering an implicit formulation, the numerical scheme through characteristic lines for (29) may be expressed by

$$(1 + \alpha \Delta t) c^{n+1} - \Delta t \nabla \cdot \left(\mathbf{K} \cdot \nabla \cdot c^{n+1} \right) = c^n - \Delta t \nabla \cdot (\mathbf{u} \cdot c^n) + \frac{\Delta t^2}{2} \sum_{i=1}^{2} \sum_{j=1}^{2} u_i u_j \frac{\partial^2 c^n}{\partial x_i \partial x_j}$$

(30)

Applying the finite element method to (30) with consistent integration in all of their terms, we obtain the following equivalent finite difference scheme

$$\frac{h^2}{12} (1 + \alpha \Delta t) \left(6 c_0^{n+1} + \sum_{i=1}^{N} c_i^{n+1} \right) + \Delta t K \left(4 c_0^{n+1} - c_1^{n+1} - c_3^{n+1} - c_4^{n+1} - c_6^{n+1} \right) =$$

$$= \frac{h^2}{12} \left(6 c_0^n + \sum_{i=1}^{N} c_i^n \right) - \Delta t \frac{h}{6} [u_1 \left(2 \left(c_1^n - c_4^n \right) + c_2^n - c_3^n + c_6^n - c_5^n \right) +$$

$$+ u_2 \left(2 \left(c_3^n - c_6^n \right) + c_2^n - c_1^n + c_4^n - c_5^n \right)] -$$

$$- \frac{\Delta t^2}{2} [u_1^2 \left(2 c_0^n - c_1^n - c_4^n \right) + u_2^2 \left(2 c_0^n - c_3^n - c_6^n \right) +$$

$$+ u_1 u_2 \left(c_1^n - c_2^n + c_3^n + c_4^n - c_5^n + c_6^n - 2 c_0^n \right)],$$

where c_i is the numerical solution at the ith node and h the parameter of the discretization. Let ν be the angle between the velocity vector and the x-axis and

$$\theta_1 = \cos\nu \qquad \theta_2 = \sin\nu$$
$$\zeta_i = \cos\left(\xi_{im}h\right) \qquad \zeta_{12} = \cos\left(\xi_{1m} + \xi_{2m}\right)h$$
$$\chi_i = \sin\left(\xi_{im}h\right) \qquad \chi_i = \sin\left(\xi_{1m} + \xi_{2m}\right)h.$$

Then the amplification factor for the numerical solution becomes

$$G(\xi_m) = \frac{\frac{1}{6}\left(3 + \zeta_1 + \zeta_2 + \zeta_{12}\right) - C^2\left[1 - \theta_1^2\zeta_1 - \theta_2^2\zeta_2 + \theta_1\theta_2\left(\zeta_1 + \zeta_2 - \zeta_{12} - 1\right)\right]}{\frac{1}{6}\left(1 - \alpha\Delta t\right)\left(3 + \zeta_1 + \zeta_2 + \zeta_{12}\right) + \frac{2C}{\mathrm{Pe}}\left(2 - \zeta_1 - \zeta_2\right)} -$$
$$-i \cdot \frac{C}{3}\frac{\theta_1\left(2\chi_1 + \chi_{12} - \chi_2\right) + \theta_2\left(2\chi_2 + \chi_{12} - \chi_1\right)}{\frac{1}{6}\left(1 - \alpha\Delta t\right)\left(3 + \zeta_1 + \zeta_2 + \zeta_{12}\right) + \frac{2C}{\mathrm{Pe}}\left(2 - \zeta_1 - \zeta_2\right)}$$

where C and Pe are the local Courant and Peclet dimensionless numbers, expressed by

$$C = \frac{|\boldsymbol{u}|\,\Delta t}{h} \qquad \mathrm{Pe} = \frac{|\boldsymbol{u}|\,h}{K},$$

where h is the characteristic dimension of the element in the flow direction with velocity $\boldsymbol{u}$.

We use the Von Neumann stability criterion to establish the stability limit on the Courant number. The most restrictive situation corresponds to the particular case in which $\nu = -\pi/4$, as shown in Table 2, where stability limits for two important values of ν are compared.

Table 2. Stability limit on Courant number for various wave vector directions and consistent integration in all terms.

Wave vector	Stability limit on Courant number
$\boldsymbol{\xi} = \xi \cdot \boldsymbol{i}$	$\dfrac{1}{\cos^2 v}\left[\dfrac{1}{\mathrm{Pe}} + \sqrt{\dfrac{1}{\mathrm{Pe}^2} + \dfrac{\cos^2 v}{6}(2 + \alpha\Delta t)}\,\right]$
$\boldsymbol{\xi} = \xi \cdot \boldsymbol{i} - \xi \cdot \boldsymbol{j}$	$\dfrac{1}{1 - \sin 2v}\left[\dfrac{2}{\mathrm{Pe}} + \sqrt{\dfrac{4}{\mathrm{Pe}^2} + \dfrac{1 - \sin 2v}{6}(2 + \alpha\Delta t)}\,\right]$

For $\alpha > 0$, there is an additional condition for stability: $\Delta t \leq 2/\alpha$. These results agree with the conclusions of Peraire, Zienkiewicz and Morgan [9], in the 1D problem and also with the result obtained in 2D by Montenegro et al. [15].

For implicit formulation with reduced integration on the c^{n+1} and c^n terms, the application of the Von Neumann criterion leads us to Table 3.

Table 3. Stability limit on Courant number for various wave vector directions and reduced integration in all terms.

Wave vector	Stability limit on Courant number
$\boldsymbol{\xi} = \xi \cdot \boldsymbol{i}$	$\dfrac{1}{\cos^2 v}\left[\dfrac{1}{\text{Pe}} + \sqrt{\dfrac{1}{\text{Pe}^2} + \dfrac{\cos^2 v}{2}(2 + \alpha\Delta t)}\right]$
$\boldsymbol{\xi} = \xi \cdot \boldsymbol{i} - \xi \cdot \boldsymbol{j}$	$\dfrac{1}{1 - \sin 2v}\left[\dfrac{2}{\text{Pe}} + \sqrt{\dfrac{4}{\text{Pe}^2} + \dfrac{1 - \sin 2v}{2}(2 + \alpha\Delta t)}\right]$

From the point of view of accuracy, this scheme is first order of numerical consistency and second order of consistency for the pure convection problem with $K = 0$.

Taylor–Galerkin scheme: The stability analysis of the Taylor–Galerkin scheme is made for the unidimensional problem, which is formulated by the implicit expression

$$\left(1 - \alpha^2\frac{\Delta t^2}{2}\theta\right)c^{n+1} - \left(\Delta t K + \frac{\Delta t^2}{2}u^2 - \alpha\Delta t^2 K(1 - \theta)\right)\theta\frac{\partial^2 c^{n+1}}{\partial x^2} -$$
$$-\alpha\Delta t^2 u\theta\frac{\partial c^{n+1}}{\partial x} = \left(1 - \alpha\Delta t + \alpha^2\frac{\Delta t^2}{2}(1 - \theta)\right)c^n +$$
$$+ \left(\Delta t K + \frac{\Delta t^2}{2}u^2 - \alpha\Delta t^2 K(1 - \theta)\right)(1 - \theta)\frac{\partial^2 c^n}{\partial x^2} -$$
$$- \left(u\Delta t - \alpha u\Delta t^2(1 - \theta)\right)\frac{\partial c^n}{\partial x}.$$

In the case of considering reduced integration on the c^{n+1} and c^n terms, the numerical amplification factor is

$$G(\xi_m) = \frac{1 - A(1 - \theta)\sin^2\dfrac{\xi_m h}{2} - i \cdot B\sin\xi_m h}{1 + A\theta\sin^2\dfrac{\xi_m h}{2} - i \cdot C\sin\xi_m h}, \tag{31}$$

where

$$A = \frac{4}{h^2}\left(K\Delta t + \frac{\Delta t^2}{2}v^2 + \alpha K\Delta t^2\right) + \alpha^2\frac{\Delta t^2}{2} \tag{32}$$

$$B = \frac{1}{h}\left(v\Delta t + v\alpha\Delta t^2(1 - \theta)\right) \tag{33}$$

$$C = \frac{1}{h}\alpha v\Delta t^2\theta \tag{34}$$

and the most restrictive case corresponds to $\xi_m h = \pi$, and then the above expression yields

$$G(\xi_m) = \frac{1 - A(1 - \theta)}{1 + A\theta}. \tag{35}$$

When the Von Neumann criterion is applied to (35), the above scheme is unconditionally stable (and hence convergent) for $\theta \geq 1/2$.

The same condition is obtained when consistent integration is considered, which corresponds to a numerical amplification factor given by

$$G(\xi_m) = \frac{1 - \left(A'(1 - \theta) + \dfrac{1}{6}\right)\sin^2 \dfrac{\xi_m h}{2} - i \cdot B \sin \xi_m h}{1 + \left(A'\theta - \dfrac{1}{6}\right)\sin^2 \dfrac{\xi_m h}{2} - i \cdot C \sin \xi_m h}, \tag{36}$$

where $A' = A + \frac{1}{12}\alpha^2 \Delta t^2$, with A defined by (32). The most restrictive condition occurs when $\xi_m h = \pi$. In this situation, the scheme becomes stable if the following conditions (37) and (38) are satisfied.

$$A' \geq 0 \tag{37}$$
$$A'(1 - 2\theta) \leq 5/3. \tag{38}$$

For values $\theta \geq 1/2$, the previous conditions are always satisfied, so we can conclude that this scheme is unconditionally stable for $\theta \geq 1/2$. From the point of view of the accuracy, this scheme is second order of numerical consistency and third order of consistency for the semiimplicit formulation with $\theta = 1/3$.

4   Numerical application

A numerical application of our model in a region of Lanzarote (Canary Islands) is presented. The computational domain has 12 Km from west to east, 17 Km from south to north and 2500 m over the maximum altitude on terrain, as shown in Fig. 3, and it has been discretized using tetrahedral elements with higher levels of discretization near the ground surface and around the emission source. The Delaunay triangulation algorithm (see [12]) has been used for generating the computational mesh, shown in Fig. 4.

4.1   Wind field adjustment

Five stations of measurement of the horizontal velocities (see Table 4) were available for this application together with sounding data, which led us to consider a slightly unstability atmospheric condition according to Pasquill stability class theory. In the upper atmosphere, the dominant direction of the wind is observed from soundings, together with the corresponding geostrophic wind speed. Data available for the simulation include:

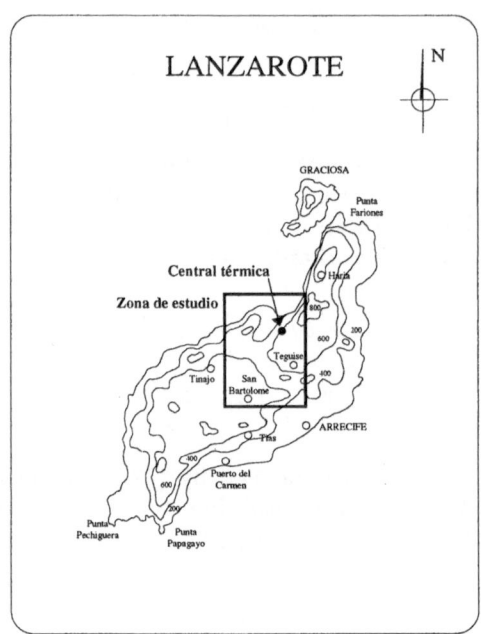

Fig. 3. Location of the computational domain in Lanzarote island.

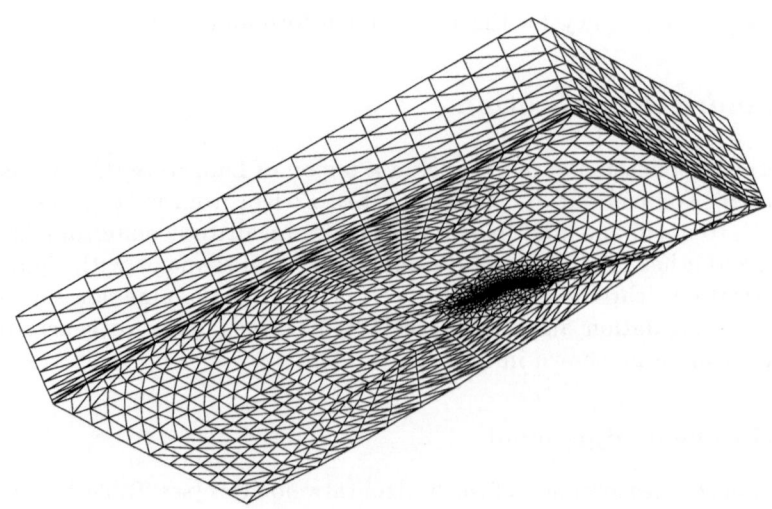

Fig. 4. Computational mesh for numerical application.

- Geostrophic wind: $\boldsymbol{u_g} = (-28.0, 28.0) \, \mathrm{m \cdot s^{-1}}$.
- Parameters of ponderation: $\tau_h = 1; \; \tau_v = 0.17$.
- Other data: $\varepsilon = 0.75; \; x_{3_0} = 0.1 \, \mathrm{m}$.

Table 4. Data from measurement stations.

Station	$X(\mathrm{Km})$	$Y(\mathrm{Km})$	$V_x(\mathrm{m \cdot s^{-1}})$	$V_y(\mathrm{m \cdot s^{-1}})$
Famara I	7,674	13,245	−3.33	−8.05
Famara II	6,956	12,053	−6.48	−6.48
Famara III	5,973	12,684	−3.30	−9.23
Famara IV	7,583	10,813	−0.84	−6.85
Tao	3,057	4,097	−0.85	−9.76

Fig. 5 shows the wind field obtained with an MMC model at different heights: 200, 270, 350, and 800 m over the sea level. From its observation, we can see:

- Dominant effect of the orography on the wind field near the ground.
- Trend of the velocity vector to raise up the orography obstacles, instead of rounding them up horizontally, because of the slight unstability considered in the computation.

4.2   Dispersion of atmospheric pollutants

We consider an emission source corresponding to a power plant, located at 8.0 Km from east side and 10.5 Km from the south limit, with a height of emission of 70 m over the terrain surface, which emits SO_2 at a uniform rate of $2.0 \, \mathrm{kg \cdot s^{-1}}$. Table 5 includes the remaining data for this application.

Table 5. Data used for the application of dispersion of pollutants.

Parameter	SO_2	H_2SO_4
$c_0(\mathrm{g \cdot m^{-3}})$	0.0	0.0
$c_2(\mathrm{g \cdot m^{-3}})$	0.0	0.0
$v_d \, (\mathrm{m \cdot s^{-1}})$	0.0044	0.0026
$v_w \, (\mathrm{m \cdot s^{-1}})$	0.28	0.14
$K_h \, (\mathrm{m^2 \cdot s^{-1}})$	25	25
$K_v \, (\mathrm{m^2 \cdot s^{-1}})$	50	50
$\alpha \, (\mathrm{s^{-1}})$	−0.0012	0.0012

In Figs. 6, 7, and 8, isolines corresponding to distribution of concentration levels of SO_2 and H_2SO_4 at steady state are shown at 200, 270, and 350 m respectively over the sea level.

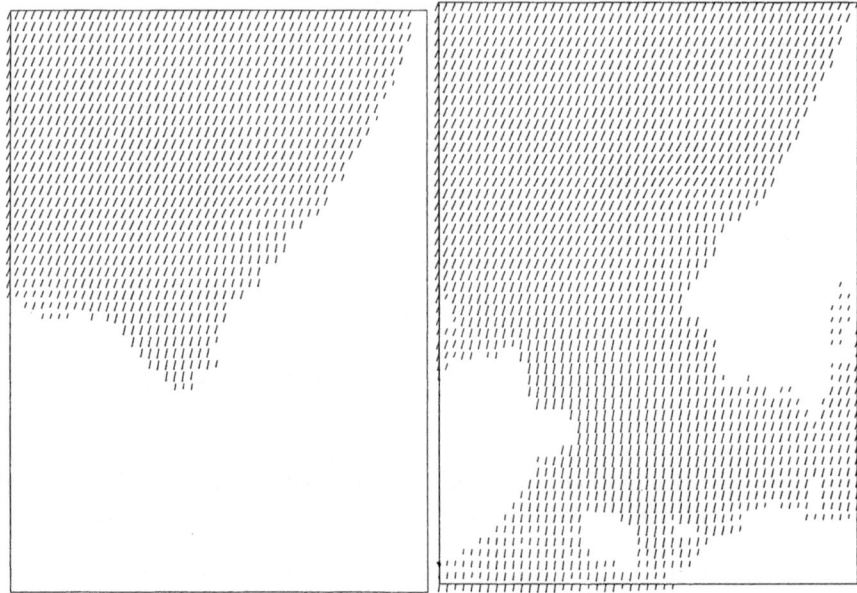

(a) Velocity vector at 200 m over the sea level

(b) Velocity vector at 270 m over the sea level

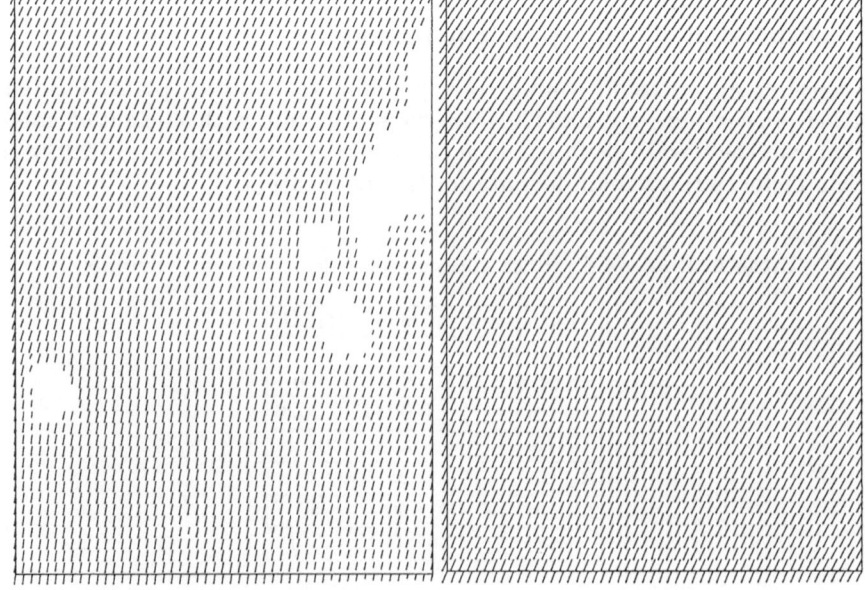

(c) Velocity vector at 350 m over the sea level

(d) Velocity vector at 800 m over the sea level

Fig. 5. Velocity field at different heights over the sea level.

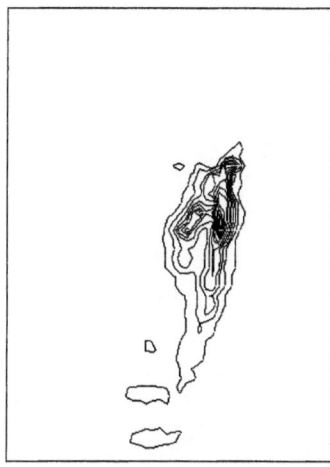
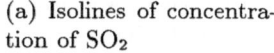

(a) Isolines of concentration of SO$_2$

(b) Isolines of concentration of H$_2$SO$_4$

Fig. 6. Isolines of concentration of pollutants at 200 m over the sea level.

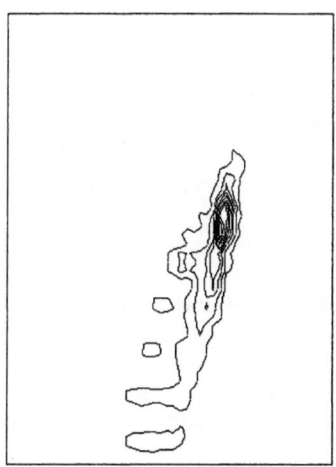
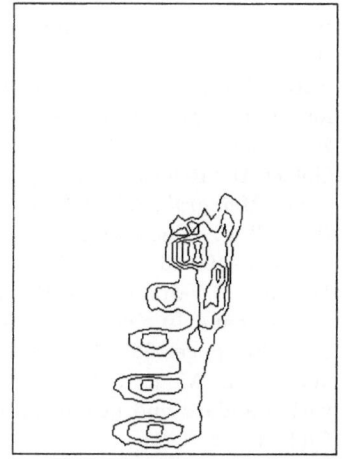

(a) Isolines of concentration of SO$_2$

(b) Isolines of concentration of H$_2$SO$_4$

Fig. 7. Isolines of concentration of pollutants at 270 m over the sea level.

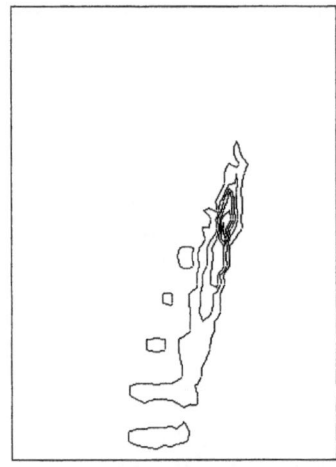

(a) Isolines of concentration of SO_2

(b) Isolines of concentration of H_2SO_4

Fig. 8. Isolines of concentration of pollutants at 350 m over the sea level.

References

1. Blackadar A. K. *Turbulence and Diffusion in the Atmosphere.* Springer-Verlag, Berlin, 1997.
2. Ratto C. F., Festa R., Romeo C., Frumento O. A., and Galluzi M. Mass consistent for wind fields over comple terrain: The state of the art. *Environ. Software,* 9:247–268, 1994.
3. Golder D. Relations among stability parameters in the surface layer. *Boundary Layer Meteorol.,* 3:47–58, 1972.
4. Pasquill F. The estimation of the dispersion of windborne material. *Meteorol. Magazine,* 90:33–49, 1961.
5. Montero G., Montenegro R., and Escobar J. M. A 3-d diagnostic model for wind adjustment. In *2nd European & African Conference on Wind Enginering: 2nd EACWE,* pages 325–332, 1997.
6. Winter G., Montero G., Ferragut L., and Montenegro R. Adaptative strategies with standard and mixed finite elements for wind field adjustment. *Solar Energy,* 54(1):49–56, 1995.
7. Panofsky H. A. and Dutton J. A. *Atmospheric Turbulence.* John Wiley, New York, 1984.
8. Palomino I. and Martin F. A simple method for spatial interpolation of the wind in complex terrain. *J. Appl. Meteorol.,* 34:1678–1693, 1995.
9. Peraire J., Zienkiewicz O. C., and Morgan K. Shallow water problems: A general explicit formulation. *Int. J. Numer. Meth. Engng.,* 22:547–574, 1986.
10. Seinfeld J. H. *Atmospheric Chemistry and Physics of Air Pollution.* Wiley Intersciences, New York, 1 edition, 1986.

11. Mata L. J., García R., and Santana R. Simulating acid deposition in tropical regions. In Baldasano J. M., Brebbia C. A., Power H., and P. Zannetti, editors, *Air Pollution II: Pollution Control and Monitoring*, pages 59–67, Boston, 1984. Computational Mechanics Publications.

12. Mücke E. P. A robust implementation for three-dimensional Delaunay triangulations. *International Journal of Computational Geometry & Applications*, 8(2):255–276, 1998.

13. Moussiopoulos N., Flassak Th., and Knittel G. A refined diagnostic wind model. *Environ. Software*, 3:85–94, 1988.

14. Benoit R. On the integral of the surface layer profile-gradient functions. *J. Appl. Meteorol.*, 16:859–860, 1977.

15. Montenegro R. *Aplicación de Métodos de Elementos Finitos Adaptativos a Problemas de Convección-Difusión*. PhD thesis, Universidad Politécnica de Canarias, 1989.

Numerical methods in oceanic circulation

Part B

Numerical methods in ocean circulation

Eulerian versus semi-Lagrangian schemes in some ocean circulation problems: a preliminary study

Rodolfo Bermejo

Universidad Complutense de Madrid
Departamento de Matemática Aplicada
28040 Madrid, Spain
rbermejo@mat.ucm.es

Abstract. A comparative study of explicit semi-Lagrangian and Eulerian schemes is carried out in the context of ocean circulation problems. We propose the explicit semi-Lagrangian schemes to overcome some computational difficulties possessed by the standard implicit semi-Lagrangian ones when they are used in ocean general circulation models formulated in spherical coordinates. The numerical comparative study of the new semi-Lagrangian schemes with Eulerian schemes, which are used in many ocean models, is performed on problems whose solutions are representative of relevant ocean circulation features.

1   Introduction

The general circulation of the ocean is a blend of motions that vary on a wide range of space and time scales. Based on the magnitude of such scales we can establish the following rough hierarchy of ocean motions which are of interest to the general circulation of the ocean: (i) *Large scale motions.* This category is characterized by motions with space scales of $O(10^3$ km) and time scales extending from years to millenia. Clear examples of such ocean phenomena are the ocean gyres driven by wind, heat and matter exchanges with the atmosphere; the equatorial current system; and the western boundary currents, such as Kuroshio and the Gulf stream, whose influences on their respective basin circulations are so profound. Also, we should include the thermohaline circulation. (ii) *The mesoscale motions.* This category is characterized by a range of space scales of $O(10$ km $- 10^2$ km). Much of the energy of the ocean is concentrated in this range, because the transient marine eddies as well as the phenomena associated with the intense boundary currents occur in this range.

As for time scales, the ocean has many dynamically important time scales extending from millenia variations of the termohaline circulation down to the rapid variations in time of surface gravity waves.

This variety of scales and dynamics is so complex and rich that it represents a difficult challenge for physicists and applied mathematicians to interpret, simulate and analyse. In this paper, we shall focus on some issues concerning with the design and efficiency of the construction of numerical models to simulate the general circulation of the ocean. Specifically, we explore some applications in ocean models of the so called semi-Lagrangian schemes whose success in numerical weather prediction and atmosphere circulation models is nowadays well established.

2   The primitive equations of the general circulation of the ocean

The design of a numerical ocean general circulation model, hereafter NOGCM, is a long process that consists of a compromise among several fields of scientific knowledge. First, the physics of the ocean motions; the second one is numerical analysis that contributes to properly formulate the numerical model; the third one, but not the less important, is computational science because the computer is the tool to carry out the physical and numerical formulation of the model. The spectacular progress made in computing power during the last decade has made possible to carry out better and more complete simulations with old numerical models such as the Cox–Bryan model. Its is only through interactions of the above three fields that efficient and accurate models can be constructed and executed.

The physics is mathematically represented by the so called primitive equations (PEs) in a spherical coordinate system on an ocean domain. The ocean is assumed to be a slightly compressible Newtonian fluid under the influence of Coriolis force. The quantities that describe the ocean circulation are velocity, pressure, temperature, salinity and density of sea water. The governing equations are formulated in a spherical coordinate system and consist of the Navier–Stokes equations for velocity and pressure, transport-diffusion equations for temperature and salinity plus one equation of state. On the basis of relevant features of the ocean circulation, the formulation of the PEs can be simplified through the following assumptions:

A1 *Boussinesq approximation.* The density of sea water is assumed to be constant except in the buoyancy terms and in the equation of state.

A2 *Thin layer approximation.* $\overline{H}/a \ll 1$, where a and $\overline{H}$ are the radius of the Earth and the maximum depth of the ocean respectively. Let (θ, φ, r) be the spherical coordinate system with the origin at the center of the Earth, where θ is the colatitude, $0 \leq \theta \leq \pi$; φ is the longitude, $0 \leq \varphi \leq 2\pi$; r is the radial component and $z = r - a$ is the vertical component with respect to the sea level. By virtue of A2 we have $z \ll a$, so we can replace r by a in the equations and substitute $\partial/\partial r$ by $\partial/\partial z$. Hence the coordinate system (θ, φ, r) becomes (θ, φ, z) with scale factor $h_1 = a$, $h_2 = a\sin\theta$ and $h_3 = 1$. The unit vectors in the θ-, φ-, and z-directions are:

$$e_\theta = \frac{1}{a}\frac{\partial}{\partial\theta}, \quad e_\varphi = \frac{1}{a\sin\theta}\frac{\partial}{\partial\varphi}, \quad e_z = \frac{\partial}{\partial z}.$$

A3 *Hydrostatic approximation.* The gravity force is in balance with the vertical component of the pressure gradient in the momentum equations.

Let M be the domain occupied by the ocean with boundary $\Gamma = \Gamma_u \cup \Gamma_b \cup \Gamma_l$, where Γ_u denotes the sea surface; Γ_b means the ocean bottom $\Gamma_b \equiv -H(\theta, \varphi)$, such that we assume that there is $H_0 > 0$ satisfying $-H(\theta, \varphi) < -H_0$; and Γ_l is the lateral boundary.

We decompose the velocity vector $\mathbf{V_3}$ as

$$\mathbf{V_3} = \mathbf{V} + w e_z, \tag{1}$$

with

$$\mathbf{V} = u e_\theta + v e_\varphi, \tag{2}$$

where

$$u = a\frac{d\theta}{dt}, \quad v = a\sin\theta\frac{d\varphi}{dt}, \quad w = \frac{dz}{dt}. \tag{3}$$

Using A1–A3, the PEs for the general circulation of the ocean are:

$$\begin{cases} \dfrac{D\mathbf{V}}{Dt} + f e_z \wedge \mathbf{V} = -\dfrac{1}{\rho_0}\nabla_\tau P + \mathbf{F}, \\[2mm] \dfrac{\partial P}{\partial z} = -\rho g, \\[2mm] \dfrac{\partial w}{\partial z} + \nabla_\tau \cdot \mathbf{V} = 0. \end{cases} \tag{4}$$

$$\frac{DT}{Dt} = \nabla_\tau \cdot (\mu_T \nabla_\tau T) + \frac{\partial}{\partial z}\left(k_T \frac{\partial T}{\partial z}\right), \tag{5}$$

$$\frac{DS}{Dt} = \nabla_\tau \cdot (\mu_S \nabla_T) + \frac{\partial}{\partial z}\left(k_S \frac{\partial S}{\partial z}\right), \tag{6}$$

$$\rho = \rho(T, S),$$

plus boundary and initial conditions for $\mathbf{V}, w, T$ and S.

In the above equations, ρ_0 is a constant reference density, f denotes the Coriolis parameter given as $f = 2|\Omega|\sin\theta$, where Ω is the angular velocity of the rotating Earth. $\mathbf{F}$ are frictional terms, whereas μ_T, μ_S and k_T, k_S are the horizontal and vertical eddy diffusion coefficients for temperature and salinity respectively. The material derivatives for $\mathbf{V}, T$ and S have the following expressions:

(i) $\dfrac{D\mathbf{V}}{Dt} = \dfrac{\partial \mathbf{V}}{\partial t} + \mathbf{V} \cdot \nabla_\tau \mathbf{V} + w\dfrac{\partial \mathbf{V}}{\partial z},$

(ii) $\dfrac{D}{Dt} = \dfrac{\partial}{\partial t} + \mathbf{V} \cdot \nabla_\tau + w\dfrac{\partial}{\partial z},$ for T and S, where

$$\nabla_\tau \mathbf{V} = \begin{bmatrix} \frac{1}{a}\frac{\partial u}{\partial \theta} & \frac{1}{a\sin\theta}\frac{\partial u}{\partial \varphi} - \frac{v}{a}\cot\theta \\[2mm] \frac{1}{a}\frac{\partial v}{\partial \theta} & \frac{1}{a\sin\theta}\frac{\partial v}{\partial \varphi} + \frac{u}{a}\cot\theta \end{bmatrix},$$

for T and S scalars,

$$\nabla_\tau T = \frac{1}{a}\frac{\partial T}{\partial \theta} + \frac{1}{a\sin\theta}\frac{\partial T}{\partial \varphi},$$

and an analogous expression for $\nabla_\tau S$. Here, ∇_τ denotes the gradient operator in the (θ, φ, z) system with respect to θ and φ. A commonly used formulation for $\mathbf{F}$ in ocean general circulation models is the following:

$$\mathbf{F} = \nabla \cdot \sigma \tag{7}$$

where σ is the symmetric stress tensor. By virtue of A3 and the transverse isotropy of the ocean motions [8] the components of σ are given as:

$$\sigma_{11} = A\left(\frac{1}{a}\frac{\partial u}{\partial \theta} - \frac{1}{a\sin\theta}\frac{\partial v}{\partial \varphi} - \frac{u}{a}\cot\theta\right) = -\sigma_{22},$$

$$\sigma_{12} = \sigma_{21} = A\left(\frac{1}{a\sin\theta}\frac{\partial u}{\partial \varphi} + \frac{1}{a}\frac{\partial v}{\partial \theta} - \frac{v}{a}\cot\theta\right),$$

$$\sigma_{13} = \sigma_{31} = k\frac{\partial u}{\partial z}, \quad \sigma_{23} = \sigma_{32} = k\frac{\partial v}{\partial z},$$

where A and k are horizontal and vertical eddy viscosity coefficients respectively.

Note that in the PEs the component w appears as a diagnostic variable that has to be calculated from the values of $\mathbf{V}$ via the divergence equation. Many NOGCMs take $w = 0$ at $z = 0$, this is the so called rigid lid boundary condition, so that by vertical integration of the hydrostatic and divergence equations we get:

$$W(\mathbf{V})(t, \theta, \varphi, z) = \nabla_\tau \cdot \int_z^0 \mathbf{V}(t, \theta, \varphi, z)\, dz, \tag{8}$$

$$\nabla_\tau \cdot \int_{-H}^0 \mathbf{V}\, dz' = 0, \tag{9}$$

and

$$p(t, \theta, \varphi, z) = p_s(t, \theta, \varphi) + \int_z^0 g\rho\, dz'. \tag{10}$$

Equation (8) is the prognostic equation for w and (9) represents a non-local constraint of the PEs whose Lagrange multiplier is $p_s(t, \theta, \varphi)$. This is a distinctive feature of the PEs in relation with the conventional three-dimensional Navier–Stokes equations for which the divergence equation is a local constraint.

The aproach to find the numerical solution of the PEs consists of discretizing in time and in space the equations. The discretization in space can be carried out either by finite elements or by finite differences. Although many numerical models use finite differences, however an advantage of finite elements is that one

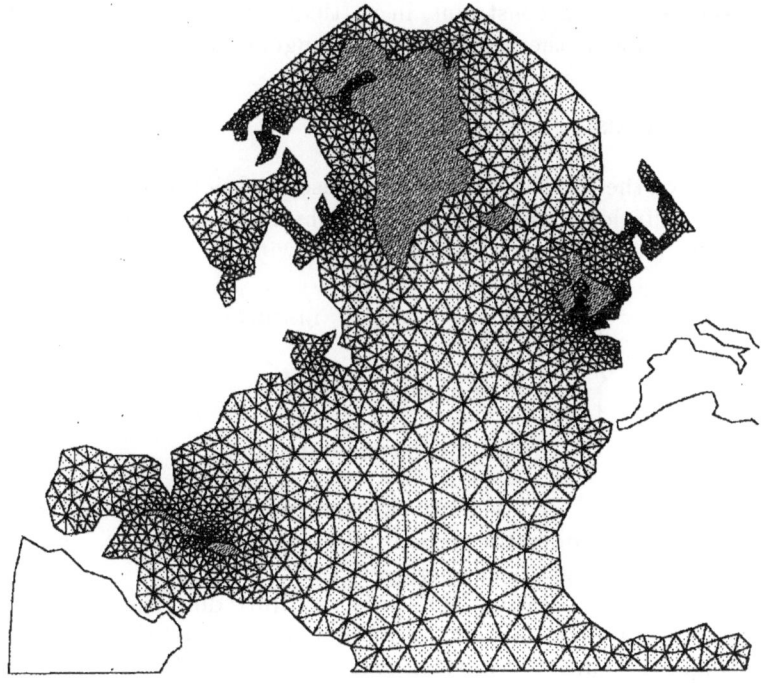

Fig. 1. A triangular finite element mesh for the North Atlantic

can refine selectively the mesh based on the dynamics and numerical properties of the solution. We show in Fig. 1 a triangular mesh for the North Atlantic Ocean generated by our finite element code still under development.

As for the time discretization, most of the NOGCMs use explicit schemes, mainly a combination of the leap-frog scheme for the advective terms with a first order explicit Euler scheme for the diffusion and friction terms. The reasons for doing so are historical and also computational. To explain the latter, let us consider a relatively small ocean domain such as the North Atlantic, on which we generate a coarse grid of about $1° \times 1°$ size with an average of 15 points in the vertical. Such a grid has $O(10^4)$ points. If we use either an implicit or semi-implicit scheme (explicit for the advective terms and implicit for the viscous and diffusion terms) to perform a long-term computation, say $O(10)$ years, we would require large amount of both computer storage and CPU time, because, in addition to the large number of unknowns, the equations for the u and v components are coupled through the non-linear and friction terms (see the expression for $\mathbf{V} \cdot \nabla_\tau \mathbf{V}$ and σ). So that ocean modelers have favored the use of explicit schemes that allow to uncoupling the u- and v- equations and require much less computer storage; however, explicit schemes have a restriction over the time step Δt since they must satisfy a stability criterium. A possible way to overcome such time step constraint, in particular if the restriction comes from the non-linear terms, is the use of semi-Lagrangian schemes.

3   The schemes

The structure of the PEs as regards time discretization can be represented by the advection-diffusion equation. Let C be a physical magnitude, where C stands for $\mathbf{V}$, T or S. The equation we are concerned with is:

$$\begin{cases} \dfrac{DC}{Dt} = \nabla \cdot (\nu \nabla C) \quad \text{in } (0,t) \times M, \\[2ex] C(0,\mathbf{x}) = C_0(\mathbf{x}), \\[2ex] BC = G \quad \text{on the boundary } \Gamma, \end{cases} \tag{11}$$

where $\frac{DC}{Dt} = \frac{\partial C}{\partial t} + \mathbf{V} \cdot \nabla C$, $\mathbf{V}$ is a velocity vector, ν is an eddy coefficient, B denotes the boundary operator, $\mathbf{x}$ is a point of the domain M, and $\nabla\cdot$ and ∇ are the divergence and gradient operators respectively.

Since most of the operating NOGCMs use finite differences with a uniform grid and a constant time step Δt, we restrict our presentation of Eulerian and semi-Lagrangian schemes to this situation. Furthermore, in order to simplify the exposition, and without loss of generality of the points we wish to make in the comparisons of these schemes, we shall assume that M is a bounded domain in $\mathbb{R}^2$ and that (11) is formulated in a Cartesian coordinate system. Thus, to compute the numerical solution of (11) we divide the internal $[0,t]$ into N subintervals $[t_n, t_{n+1}]$, $n = 0, 1, \ldots, N - 1$, of equal length Δt and define a regular grid over M with grid points $\mathbf{x}_{ij}$ such that $i = 1, \ldots, I$ and $j = 1, \ldots, J$. We use the notation $\mathbf{x}_{ij} \equiv (x_i, y_j)$ and $a_{ij}^n \equiv a\,(\mathbf{x}_{ij}, t_n)$ unless otherwise stated. We define an *Eulerian scheme* as a scheme that for all n and all indices i, j approximates $\frac{\partial F}{\partial t}$ at each grid point $\mathbf{x}_{ij}$ by a time discretization scheme. In contrast, we define a semi-Lagrangian scheme as one that for each subinterval $[t_n, t_{n+1}]$ approximates (11) along the trajectories of the fluid particles that will reach the grid points $\mathbf{x}_{ij}$ at time t_{n+1} [6].

3.1   An Eulerian scheme

A very much used explicit Euler scheme in NOGCMs is the following [1]:
  For all $n = 1, 2, \ldots, N - 1$, $i = 1, \ldots, I$, and $j = 1, \ldots, J$, compute

$$\frac{C_{ij}^{n+1} - C_{ij}^{n-1}}{2\Delta t} + (\mathbf{V} \cdot \nabla_h C)_{ij}^n = (\nabla_h \cdot (\nu \nabla C))_{ij}^{n-1}. \tag{12}$$

Here the operators $\nabla_h \cdot$ and ∇_h are the finite difference discretizations for $\nabla\cdot$ and ∇, respectively. The scheme (12) combines the leap-frog scheme for advection with the explicit Euler scheme for diffusion/dissipation. Relevant properties of this scheme are:

(i) The scheme is easy to implement and well adapted to any type of machine, i.e., scalar, vector or parallel machines.

(ii) The scheme keeps the storage requirements reasonably low.

(iii) It has a truncation error $O(\Delta x^2) + O(\Delta t)$.

(iv) It is conditionally stable and has to satisfy the following restrictive criterium:

$$2 \left(\frac{\Delta t \max|\mathbf{V}|}{\Delta x} \right)^2 + \left(\frac{4 d \mu \Delta t}{\Delta x^2} \right) < 1,$$

where d is the dimension of space ($d = 1$, 2 or 3). Consequently, for fine grids or grids with refinement the time step of this scheme is very small. The scheme is used for most of the NOGCMs based on the Cox–Bryan model.

3.2   An explicit semi-Lagrangian Runge–Kutta scheme

Semi-Lagrangian schemes have shown to be very efficient in atmospheric models; however, they have not been used very much in ocean modeling. The way these schemes are implemented in meteorology or atmospheric pollution models combines the integration of the advective terms along the trajectories of the fluid particles with an implicit formulation for the diffusive/viscous terms. As we saw in Section 2, the viscous terms of the u and v components are coupled, so the use of implicit formulation of these terms, even within a semi-Lagrangian framework, may not be affordable. Therefore, we shall describe and test semi-Lagrangian schemes that deal with the diffusion terms of (10) in the same way as explicit Runge–Kutta methods do.

In order to facilitate the presentation of our explicit semi-Lagrangian scheme, let us briefly recall the formulation of the first and second order explicit Runge–Kutta schemes as they are used to find the numerical solution to initial value problems. To this end, let us consider the system:

$$\begin{cases} \dfrac{d\mathbf{z}}{dt} = \mathbf{f}(t, \mathbf{z}), \quad t_0 < t \leq T, \\[2mm] \mathbf{z}(t_0) = \mathbf{z}_0, \end{cases} \tag{13}$$

where $\mathbf{z}$ and $\mathbf{f}$ are in $\mathbb{R}^m$, $m \geq 1$. We approximate the solution to (13) by the following second order Runge–Kutta method:

Let $\mathbf{z}^0 = \mathbf{z}_0$. For $n = 0, 1, \ldots, N - 1$, compute

$$\mathbf{K}_1 = \Delta t \, \mathbf{f}\left(t_n, z^n \right),$$

$$\mathbf{K}_2 = \Delta t \, \mathbf{f}\left(t_n + \tfrac{1}{2}\Delta t, \, z^n + \tfrac{1}{2}\mathbf{K}_1 \right),$$

and set

$$\mathbf{z}^{n+1} = \mathbf{z}^n + \mathbf{K}_2. \tag{14}$$

If in (14) we put

$$\mathbf{z}^{n+1} = \mathbf{z}^n + \mathbf{K}_1, \tag{15}$$

then we obtain the first order explicit Euler scheme.

To construct the semi-Lagrangian Runge–Kutta schemes for computing the solution of (11) we need to calculate, in each subinterval $[t_n, t_{n+1}]$, the trajectories $\mathbf{X}_{ij}\left(\mathbf{x}_{ij}, t_{n+1}; t\right)$ of the particles that reach the grid points $\mathbf{x}_{ij}$ at t_{n+1}. This is done by solving for each pair i, j the system of equations:

$$
\begin{cases}
\dfrac{d\mathbf{X}_{ij}}{d\tau} = \mathbf{V}\left(\mathbf{X}_{ij}\left(\mathbf{x}_{ij}, t_{n+1}; \tau\right), \tau\right), \\[2mm]
\dfrac{dt}{d\tau} = 1, \quad t_n \leq \tau \leq t_{n+1}, \\[2mm]
\mathbf{X}_{ij}\left(\mathbf{x}_{ij}, t_{n+1}; t_{n+1}\right) = \mathbf{x}_{ij}.
\end{cases}
\tag{16}
$$

For $t \in [t_n, t_{n+1}]$, the differential of arc ds_{ij} of the trajectory $\mathbf{X}_{ij}\left(\mathbf{x}_{ij}, t_{n+1}; t\right)$ is given by

$$
ds_{ij} = \sqrt{dt^2 + dX_{ij}^2 + dY_{ij}^2} = dt\, \psi_{ij}(t),
$$

where, by virtue of (16), $\psi_{ij}(t)$ is expressed as

$$
\psi_{ij}(t) = \sqrt{1 + u^2\left(\mathbf{X}_{ij}\left(\mathbf{x}_{ij}, t_{n+1}; t\right)\right) + v^2\left(\mathbf{X}_{ij}\left(\mathbf{x}_{ij}, t_{n+1}; t\right)\right)}.
$$

Thus,

$$
s_{ij}(t) = \int_{t_n}^{t} \psi_{ij}(\tau)\, d\tau.
$$

Note that $s_{ij}\left(t_n\right) = 0$, and $s_{ij}\left(t_{n+1}\right)$ is the arc length from the departure point $\mathbf{X}_{ij}\left(\mathbf{x}_{ij}, t_{n+1}; t_n\right)$ to the grid point $\mathbf{x}_{ij}$ at t_{n+1}.

Let us denote by $C_{ij}\left(s(t)\right)$ the value of $C(\mathbf{x}, t)$ associated with the particle that describes the trajectory $\mathbf{X}_{ij}\left(\mathbf{x}_{ij}, t_{n+1}; t\right)$. Then it follows that, for $t_n \leq t \leq t_{n+1}$,

$$
\frac{dC_{ij}}{ds} = \frac{1}{\psi_{ij}(t)} \frac{DC_{ij}}{Dt}.
$$

Hence, we approximate (11) by the semi-discrete system of ordinary differential equations:

$$
\begin{cases}
\dfrac{dC_{ij}}{ds} = \dfrac{1}{\psi_{ij}(t)} \nabla_h \cdot \left(\nu \nabla_h C_{ij}\right), \quad t_n \leq \tau \leq t_{n+1}, \\[2mm]
C_{ij}\left(s(t_n)\right) = C_{ij}^n\left(\mathbf{X}_{ij}\left(\mathbf{x}_{ij}, t_{n+1}; t_n\right)\right), \\[2mm]
BC_{ij}\left(s(t)\right) = g_{ij} \text{ on the boundary } \Gamma.
\end{cases}
\tag{17}
$$

Based on the formulation of the conventional Runge–Kutta explicit schemes introduced above, we are now in a position to formulate the explicit semi-Lagrangian Runge–Kutta schemes to integrate (17) as follows:

(i) For all grid points $\mathbf{x}_{ij}$ find $\mathbf{X}_{ij}\left(\mathbf{x}_{ij}, t_{n+1}; t_n\right)$ by solving (16) using, for instance, scheme 2 of Temperton and Staniforth [7].

(ii) Evaluate $C_{ij}^{*n} \equiv Cij^n \left(\mathbf{X}_{ij}\left(\mathbf{x}_{ij}, t_{n+1}; t_n\right)\right)$ by an interpolation procedure from the values $\tilde{C}ij^n$. The interpolation procedure we have employed in the numerical tests of this paper are piecewise bicubic or biquadratic Lagrange interpolation.

For each pair i, j, **DO:**

(iii) Set

$$\Delta s = \Delta t\, \psi_{ij}$$

and then evaluate

$$K_{1ij} = \Delta t\, \nabla_h \left(\nu \nabla_h C_{ij}^{*n}\right).$$

(iv) Set

$$\overline{C}_{ij} = C_{ij}^{*n} + \tfrac{1}{2} K_{1ij}$$

and compute K_{2ij} as

$$K_{2ij} = \Delta t\, \nabla_h \left(\nu \nabla_h \overline{C}_{ij}\right).$$

(v) Then compute

$$C_{ij}^{n+1} = C_{ij}^{*n} + K_{2ij}$$

to obtain an explicit second order semi-Lagrangian Runge–Kutta scheme (SLRK2), or

$$C_{ij}^{n+1} = C_{ij}^{*n} + K_{1ij}$$

to obtain the explicit first order semi-Lagrangian Runge–Kutta scheme (SLRK1).

The relevant properties of these schemes are:

(i) They are explicit, so they are well adapted to run on any type of computer architecture.

(ii) SLRK2 has a truncation error $O(\Delta x^2) + O(\Delta y^2) + O(\Delta t^2)$, while the truncation error of SLRK1 is $O(\Delta x^2) + O(\Delta y^2) + O(\Delta t)$.

(iii) It is well known that semi-Lagrangian schemes are unconditionally stable for the advective terms [4]; however, SLRK1 and SLRK2 are both explicit for the diffusion terms, because of this they are conditionally stable. It is easy to prove by Von Neuman analysis of linear stability that a necessary condition for SLRK1 and SLRK2 to be stable is:

$$\frac{\nu \Delta t}{\Delta x^2} \leq \frac{1}{2d},$$

where d is the dimension of space.

64   R. Bermejo

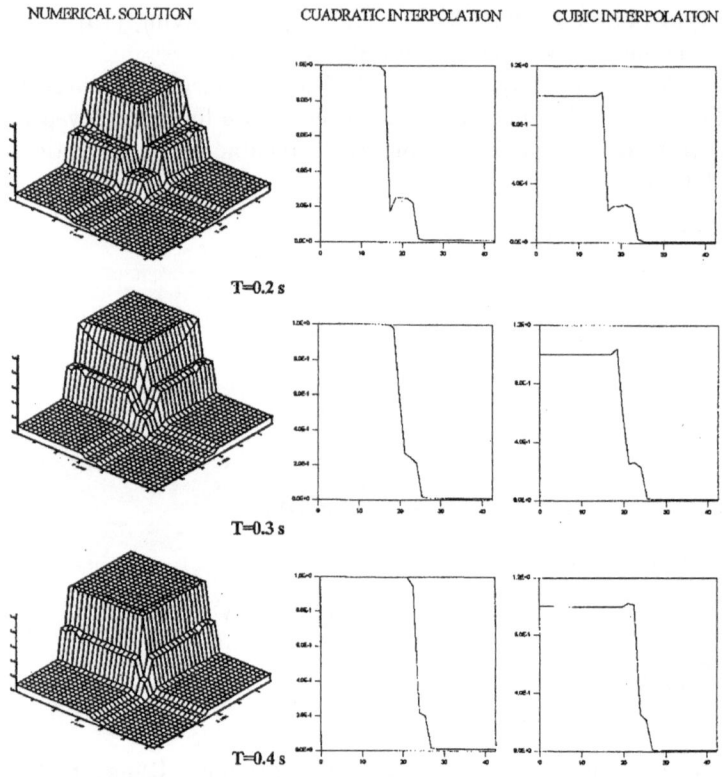

Fig. 2. Semi-Lagrangian solution at different times

4   Numerical tests

4.1   The two-fronts problem

Our first benchmark problem consists of the advection-diffusion of two fronts
that travel at different speeds and eventually coalesce into one front. The maximum velocity of both fronts is along the main diagonal of the square domain D.
This problem has been previously solved by Krishnamachari et al. [3] and represents an interesting test to illustrate the behaviour of the numerical schemes in
the numerical simulation of oceanic and atmospheric fronts. The mathematical
formulation of the problem is:

$$\frac{\partial c}{\partial t} + u(x,t)\frac{\partial c}{\partial x} + v(y,t)\frac{\partial c}{\partial y} = \nu \Delta c \quad \text{in } D \times (0,T], \tag{18}$$

$$c|_\Gamma = u(x,t)v(y,t)|_\Gamma, \tag{19}$$

$$c(\mathbf{x},0) = u(x,0)v(y,0), \tag{20}$$

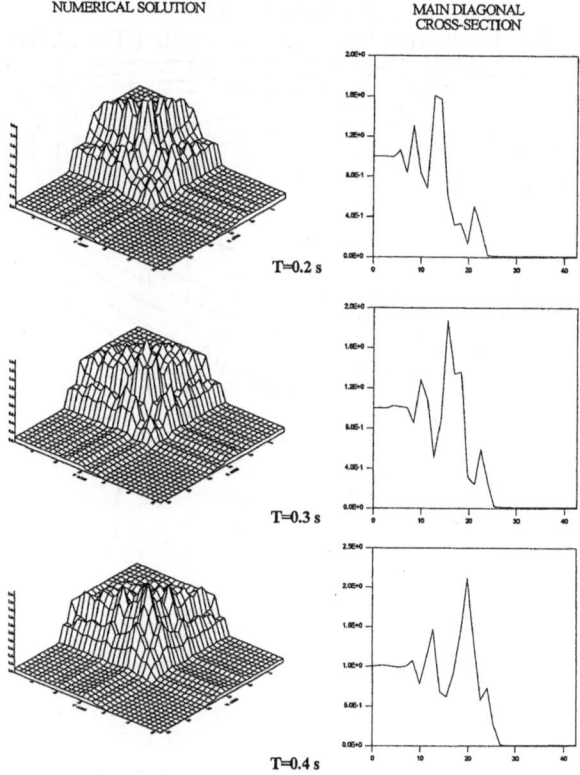

NUMERICAL SOLUTION

MAIN DIAGONAL
CROSS-SECTION

T=0.2 s

T=0.3 s

T=0.4 s

Fig. 3. Eulerian solution at different times

where

$$v_i(\xi, t) = \frac{0.1e^{-A_i} + 0.5e^{-B_i} + e^{-C_i}}{e^{-A_i} + e^{-B_i} + e^{-C_i}}, \quad i = 1, 2, \tag{21}$$

$$A_i = \frac{0.05}{\nu}(\xi - 0.5 + 4.95t), \ B_i = \frac{0.25}{\nu}(\xi - 0.5 + 0.75t), \ C_i = \frac{0.5}{\nu}(\xi - 0.375),$$

with $\xi = x$ for $i = 1$, and $\xi = y$ for $i = 2$. The analytical solution is given as

$$c(\mathbf{x}, t) = u(x, t)v(y, t). \tag{22}$$

The parameters used in this example are $\Delta x = \Delta y = \frac{1}{31}$, $\nu = 5 \times 10^{-4} \, \mathrm{m^2 s^{-1}}$, $\Delta t = 5 \times 10^{-2} \, \mathrm{s}$, and $T = 0.6 \, \mathrm{s}$. The numerical solution obtained by the SLRK2 scheme with biquadratic and bicubic Lagrange interpolation is displayed in Figure 2, where we have included three dimensional plots together with cross-sections along the main diagonal. The results of a fully implicit Crank–Nicolson Eulerian scheme are shown in Figure 3. From a simple inspection of these figures, it is clear that the semi-Lagrangian solutions are more accurate than the Eulerian ones. In order to further illustrate this fact, we compute pointwise errors

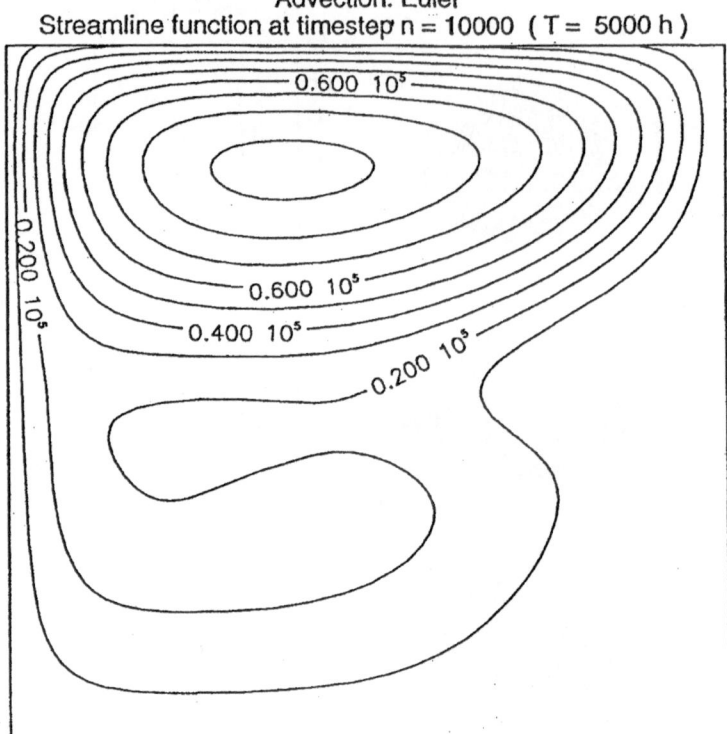

Advection: Euler
Streamline function at timestep n = 10000 (T = 5000 h)

Fig. 4.

along the main diagonal and a longitudinal cross-section for the SLRK2 with biquadratic interpolation and Eulerian schemes, respectively. We notice that in the semi-Lagrangian schemes, large errors are concentrated in narrow regions around the two steep fronts, whereas outside such regions the errors are very small. In contrast, we observe that in the Eulerian scheme, there are wakes of large errors, left behind the steep gradients, which are distributed over wide regions.

4.2   The wind driven homogeneus ocean

Our second example is the wind driven circulation of a β-plane homogeneus ocean. There are various issues we wish to address in this numerical experiment. The first one is referred to the numerical diffusion of semi-Lagrangian schemes. The reason for this is to test the validity of the extended idea in the ocean modeling community that semi-Lagrangian schemes are more diffusive

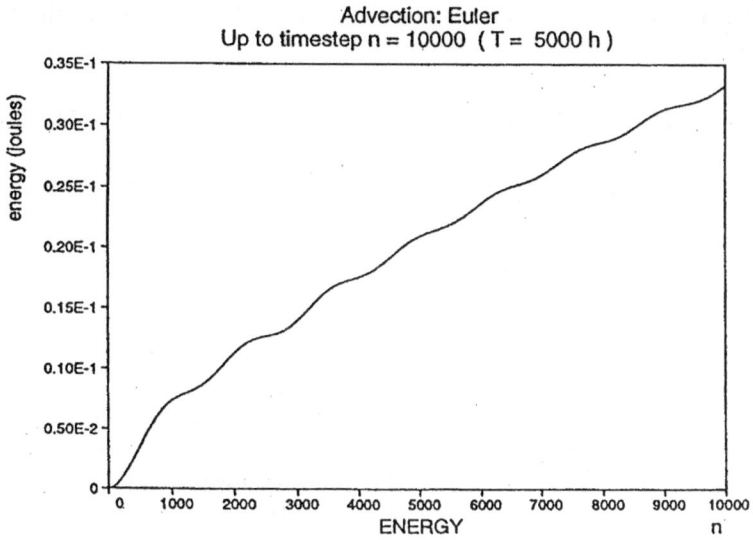

Fig. 5.

than the Eulerian schemes being currently used in ocean models; hence, semi-Lagrangian schemes are not suitable for the type of long-term computations wihich are needed in ocean studies. Of course, this is true if semi-Lagrangian schemes employ linear interpolation; however, if higher order interpolation is used then things may not be that way. So that, we want to examine the effects of the numerical diffusion of semi-Lagrangian schemes with higher order interpolation on long term simulation of a wind driven homogeneus ocean circulation. The second issue is concerned with stability, accuracy and computational cost of semi-Lagrangian schemes versus Eulerian ones in long term computations of ocean models. Our third concern has to do with the interpolation procedure used in the calculations done at the departure points. Bicubic spline interpolation is admitted as a very good interpolator, albeit expensive, but not suitable when· the domain has an irregular boundary. Other interpolator frequently used in atmosphere models is piecewise bicubic Lagrange interpolation, and this could also be implemented in ocean models. However, if one is interested in modeling with finite elements with triangular grids, then piecewise biquadratic Lagrange interpolation is a natural choice compatible with the finite element formulation. Because of these arguments, we use piecewise biquadratic Lagrange interpolation in the semi-Lagrangian schemes for the numerical simulations of this example. We must say, in advance, that the conclusions that emerge from this experiment should be considered as orientative rather than conclusive. However, it is our hope that they may contribute to clarify some misunderstandings about semi-Lagrangian schemes that are so extended in the ocean modeling community.

Our barotropic ocean model is mathematically formulated by the following equations [5]:

$$\frac{\partial w}{\partial t} + J(w, \psi) + \beta \frac{\partial \psi}{\partial x} = \nu \Delta w - \gamma w + \frac{1}{H} \operatorname{curl}_z \tau \quad \text{in } M \times (0, T], \qquad (23)$$

$$w|_\Gamma = w_b(x, y, t) \quad \text{and} \quad w(x, y, 0) = 0 \quad \text{in } M, \qquad (24)$$

$$\Delta \psi = w \quad \text{in } M \times (0, T], \qquad (25)$$

$$\psi|_\Gamma = 0, \quad \text{and} \quad \psi(x, y, 0) = 0 \quad \text{in } M, \qquad (26)$$

where $w = \frac{\partial v}{\partial x} - \frac{\partial u}{\partial y}$ is the relative vorticity, ψ is the stream function which is related to the velocity components by $u = -\frac{\partial \psi}{\partial y}$ and $v = \frac{\partial \psi}{\partial x}$. J is the Jacobian operator given by $\frac{\partial w}{\partial y} \frac{\partial \psi}{\partial x} - \frac{\partial w}{\partial x} \frac{\partial \psi}{\partial y}$, M is a square domain of side $L = 1000$ Km, β is the first order latitudinal variation of the planetary vorticity $f = f_0 + \beta \left(y - \frac{1}{2} L \right)$, f_0 being the value of f at the mid-latitude of the domain. H denotes the constant depth of the ocean and τ is the steady wind stress acting at the upper surface with components $\tau_1 = 0$, $\tau_2 = -\tau_0 \cos(\pi y / L)$. ν is a lateral eddy viscosity coefficient and γ is the bottom friction coefficient. We shall compare the semi-Lagrangian solutions of (23)–(26) with the one obtained by the Eulerian scheme previously employed by many researchers such as Holland [2] and others to compute the numerical solutions of (23)–(26). Thus, the Eulerian model is given by the following equations:

$$\alpha_1 w_{ij}^{n+1} - \Delta t \nu \Delta_h w_{ij}^{n+1} = \alpha_2 w_{ij}^{n-1} + \Delta t \nu \Delta_h w_{ij}^{n-1} - $$
$$2\Delta t \left(J_h \left(w_{ij}^n, \psi_{ij}^n \right) + \beta \delta_x \psi_{ij}^n \right) + \frac{2\Delta t}{H} \operatorname{curl}_z \tau, \qquad (27)$$

$$\Delta_h \psi_{ij}^n = w_{ij}^n, \qquad (28)$$

with homogeneus boundary conditions for both ψ and w. In equations (27) and (28), Δ_h denotes the 5-point discrete Laplace operator, δ_x is the centered difference operator in the x-direction, $\alpha_1 = (1 + \Delta t \gamma)$, $\alpha_2 = (1 - \Delta t \gamma)$ and the Jacobian operator J_h in (27) is the discrete Arakawa Jacobian. This Eulerian method is known to be $O(\Delta x^2) + O(\Delta y^2) + O(\Delta t^2)$ and has to satisfy the CFL condition because the nonlinear terms are explicit in time.

The semi-Lagrangian scheme that we use to compute the solutions of (23)–(26) is the SLRK2. By virtue of the relationship between the stream function and the velocity components, we can write (23) as an advection-diffusion equation suitable for the application of semi-Lagrangian schemes. Thus, we apply SLRK2 to the following equation:

$$\frac{Dw}{Dt} = \nu \Delta w - \gamma w - \beta \frac{\partial \psi}{\partial x} + \frac{1}{H} \operatorname{curl}_z \tau \quad \text{in } M \times (0, T]. \qquad (29)$$

The velocity components, which are needed to calculate the departure points, are evaluated by the approximations $u_{ij}^n = -\delta_y \psi_{ij}^n$ and $v_{ij}^n = \delta_x \psi_{ij}^n$, where δ_x, δ_y are centered difference operators in the x- and y- directions respectively.

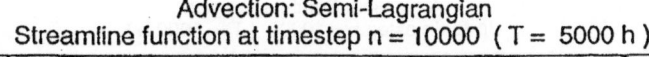

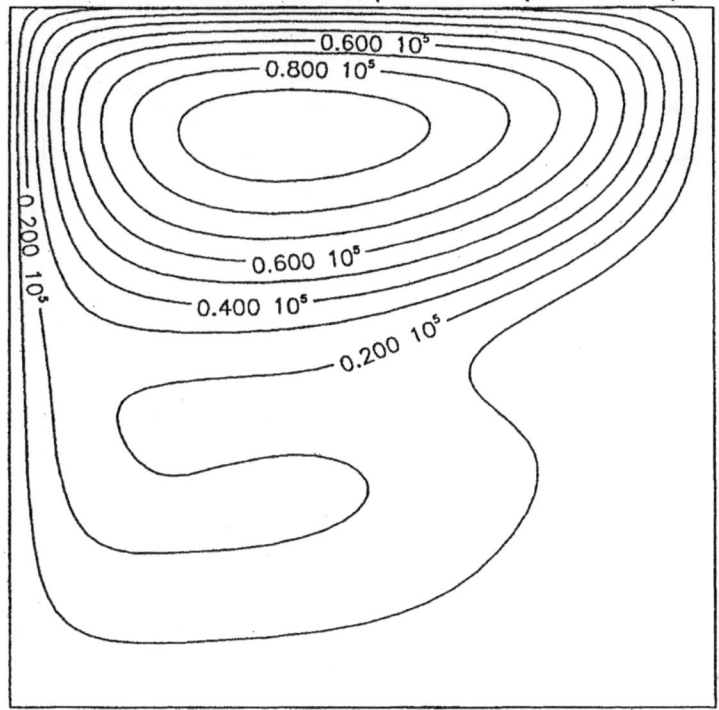

Fig. 6.

The discrete explicit SLRK2 scheme is formulated as follows:

$$
\left\{
\begin{aligned}
& k_{1ij} = \Delta t \left(\nu \Delta_h w_{ij}^{*n} - \gamma w_{ij}^{*n} \right) + \Delta t \left(\frac{1}{H} \operatorname{curl}_z \tau_{ij}^* - \beta \delta_x \psi_{ij}^{*n} \right) \\[2mm]
& \overline{w}_{ij} = w_{ij}^{*n} + \tfrac{1}{2} k_{1ij} \\[2mm]
& k_{2ij} = \Delta t \left(\nu \Delta_h \overline{w}_{ij} - \gamma \overline{w}_{ij} \right) + \Delta t \left(\frac{1}{H} \operatorname{curl}_z \tau_{ij} \left(\mathbf{X}_{ij}^{n+\frac{1}{2}} \right) - \beta \delta_x \psi_{ij}^{*n+\frac{1}{2}} \right) \\[2mm]
& w_{ij}^{n+1} = w_{ij}^{*n} + k_{2ij}
\end{aligned}
\right.
\tag{30}
$$

where we have introduced the notation $\mathbf{X}_{ij}^{n+\frac{1}{2}}$ to denote $\mathbf{X}_{ij} \left(\mathbf{x}_{ij}, t_{n+1}; t_{n+\frac{1}{2}} \right)$,

$$
\tau_{ij}^* \equiv \tau_{ij} \left(\mathbf{X}_{ij} \left(\mathbf{x}_{ij}, t_{n+1}; t_n \right) \right),
$$

$$
\psi_{ij}^{*n} \equiv \psi_{ij}^n \left(\mathbf{X}_{ij} \left(\mathbf{x}_{ij}, t_{n+1}; t_n \right) \right),
$$

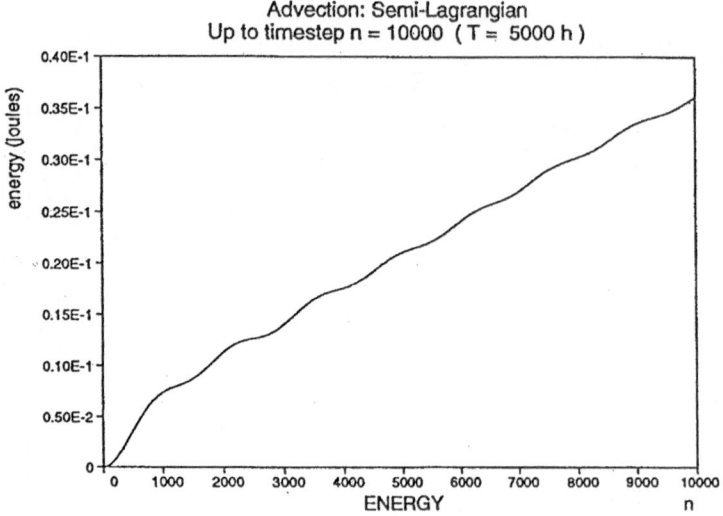

Fig. 7.

and

$$\psi_{ij}^{*n+\frac{1}{2}} \equiv \psi_{ij}^{n+\frac{1}{2}}\left(\mathbf{X}_{ij}^{n+\frac{1}{2}}\right),$$

with

$$\psi_{ij}^{*n+\frac{1}{2}} = \frac{1}{2}\left(3\psi_{ij}^{n} - \psi_{ij}^{n-1}\right).$$

The Eulerian scheme is run with parameter values: $\Delta t_1 = 1800\,\mathrm{s}$, $\Delta x = \Delta y = 20\,\mathrm{Km}$, $\beta = 2.0 \times 10^{-11}\,\mathrm{m}^{-1}\mathrm{s}^{-1}$, $\tau_0 = 2.0 \times 10^{-4}\,\mathrm{ms}^{-2}$, $H = 800\,\mathrm{m}$, $\nu = 340\,\mathrm{m}^2\mathrm{s}^{-1}$, $\gamma = 10^{-7}\,\mathrm{s}^{-1}$. Notice that the values for γ and ν were previously used by Holland [2] in some of his eddy resolving models. In the semi-Lagrangian runs we change the value of Δt_1 and maintain the values of the other parameters.

Figure 4 displays a snapshot of the streamfunction after 10000 time steps of integration ($T = 208.33$ days) of the Eulerian scheme. Figure 5 represents the total kinetic energy curve of this experiment. Figure 6 shows a streamfunction plot after 10000 steps of semi-Lagrangian integration with $\Delta t_1 = 1800\,\mathrm{s}$. The total kinetic energy curve of this semi-Lagrangian experiment is shown in Figure 7.

A simple visual comparison of these figures reveals that the semi-Lagrangian scheme, with biquadratic interpolation, gives a flow that is qualitatively similar to the one obtained by the Eulerian scheme, but with a ore intense recirculation and with a silghtly higher energy than the Eulerian model. The interesting thing to remark about these figures is that if the numerical diffusion of the semi-Lagrangian schemes had a strong influence on the long term computations, then the energy curve of the semi-Lagrangian results would have been different (with

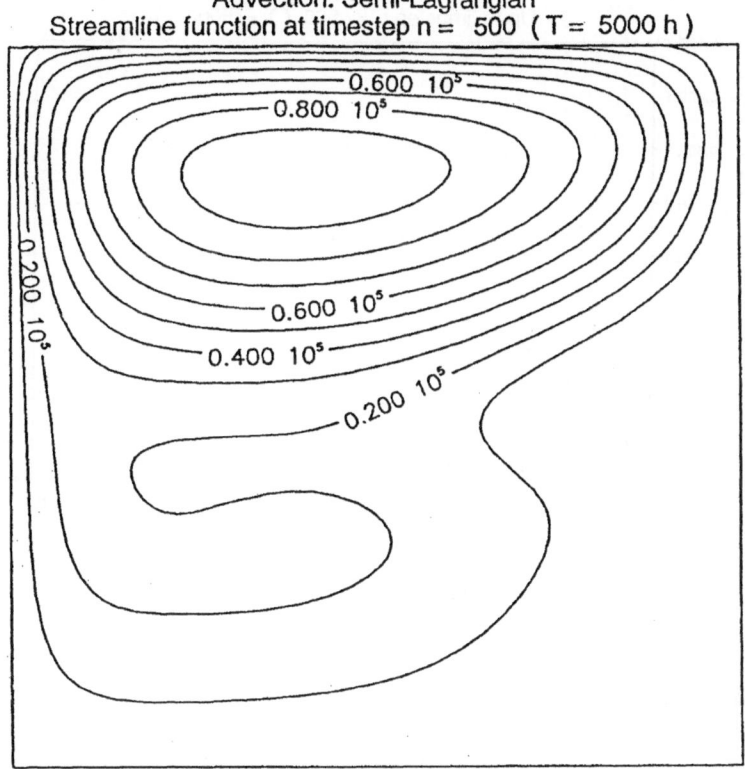

Advection: Semi-Lagrangian
Streamline function at timestep n = 500 (T = 5000 h)

Fig. 8.

lower values) from the Eulerian curve. Since this is not the case in our experiment, we may say that for sufficiently fine grids the semi-Lagrangian schemes with biquadratic interpolation are not more diffusive than Eulerian schemes.

Figures 8 and 9 show stream function plots for $T = 208.33$ days and the total kinetic energy curve of the numerical results obtained by semi-Lagrangian schemes with $\Delta t = 20\Delta t_1$. We observe in these figures that up to $\Delta t = 20\Delta t_1$ the semi-Lagrangian results are basically the same as the Eulerian ones, and it is at $\Delta t = 40\Delta t_1$ when the differences are appreciable.

In this experiment, the CPU time spent to perform a semi-Lagrangian step is about four times larger than for the Eulerian step; however, since the semi-Lagrangian schemes are able to yield numerical results as accurate as the Eulerian scheme with Δt twenty times larger, then the use of semi-Lagragian schemes can represent significant savings in computing time.

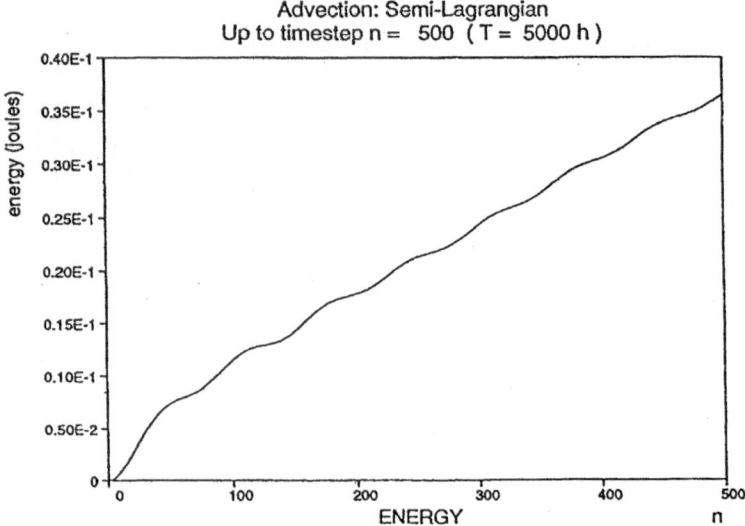

Fig. 9.

5   Conclusions

The principal conclusions reached through this preliminary comparative study are the following.

(i) For ocean models with moderate or low eddy viscosity (or diffusion) coefficients, as those used in ocean eddy resolving models, the explicit semi-Lagrangian schemes with quadratic Lagrange interpolation may be a valid alternative in the design of new numerical ocean circulation models.

(ii) For models with large eddy viscosity (or diffusion) coefficients, the restriction imposed on Δt by the stability criterium that explicit semi-Lagrangian schemes have to satisfy, may sometimes requires a Δt unreasoneable small.

(iii) For models formulated in Cartesian coordinates, or with no viscous coupling, semi-Lagrangian Crank–Nicolson schemes are a good choice because they are both accurate and unconditionally stable.

The above conclusions should be taken as orientative, rather than definitive. However, it is clear that under suitable conditions explicit semi-Lagrangian schemes show a considerable promise in ocean general circulation models.

Acknowledgements

The author was partially supported by grant CLI95-1823 from Comisión Interministerial de Ciencia y Tecnología.

References

1. Cox, M. D., 1984: GFDL Ocean Group Technical Report No. 1 (unpublished).
2. Holland, W. R., 1978: The role of mesoscale eddies in the general circulation of the ocean-numerical experiments using a wind-driven quasi-geostrophic model. J. Phys. Oceanogr. 10, 1010–1031.
3. Krisnamachari, S. V., L. J. Hayes and T. F. Russell, 1989: A finite element alternating-direction method combined with a modified method of characteristics for convection-diffusion problems. SIAM J. Numer. Anal. 26, 6, 1462–1473.
4. McDonald, A., 1984: Accuracy of multiple-upstream semi-Lagrangian advective schemes. Mon. Wea. Rev. 112, 1267–1275.
5. Pedlosky, J., 1979: Geophysical Fluid Dynamics. Springer-Verlag, 624 pp.
6. Staniforth, A. and J. Côté, 1991: Semi-Lagrangian integration schemes for the atmospheric models: a review. Mon. Wea. Rev. 113, 1050–1065.
7. Temperton, C. and A. Staniforth, 1987: An efficient two-time-level semi-Lagrangian semi-implicit integration scheme. Quart. J. Roy. Meteor. Soc. 113, 1025–1039.
8. Wajsowicz, C. R., 1993: A consistent formulation of the anisotropic stress tensor for use in models of large-scale ocean circulation. J. Comp. Phys. 105, 333–338.

Numerical simulation in oceanography. applications to the Alboran Sea and the Strait of Gibraltar

J. Macías, C. Parés, and M. J. Castro

Departamento de Análisis Matemático, Universidad de Málaga,
Campus de Teatinos s/n, 29080 Málaga, Spain
grupo@anamat.cie.uma.es
http://alboran.cie.uma.es

1 Introduction

Mathematical and numerical models are now fundamental tools in Marine Science, as in many other fields of scientific research. Nevertheless, the models, however complex, may provide at best a reasonable estimate of the behaviour of the system under study. We must be aware of this, and avoid overly relying and blindly believing in model predictions. In the task of better understanding the role that models can play in different scientific disciplines, it appears suitable to classify them according to their scope and objectives. Following *Nihoul (1994)*, three kinds of models can be listed depending on their objectives:

– *Test-oriented models* are used to test mathematical and numerical models and their implementation. These models are based on a reduced set of equations and, hence, are not intended to realistically simulate reality.

– *Process-oriented models*, which focus on a few dominant processes. This often implies sacrificing some of the realism of the results, but allows understanding the main mechanisms driving the system under study.

– *System-oriented models*, used to understand and/or predict a whole system. The results of these models must be as realistic as possible, which usually requires an appropriate data assimilation procedure to operate together with the prognostic model. Encompassing a large number of processes, system-oriented models are generally very complex, so that they are probably not the tool best suited for investigating or trying to explain the processes or mechanisms.

The long-term objective of our research focuses on the development and implementation of different numerical models for the simulation of the dynamics in the most western part of the Mediterranean Sea (the Alboran Sea and the Strait of Gibraltar), at various temporal and spatial scales. In order to do this, the initial step consists in developing some *test-oriented models* to be the base of more complex models. In this contribution we present a further stage, a *process-oriented model* that tries to represent the main characteristics of the large-scale dynamics in this region of the Mediterranean. In future research, we plan to develop another *process-oriented model* for the study of the generation of internal tides in the Strait of Gibraltar. The final goal in this wider project is to develop a *system-oriented model* for the Strait of Gibraltar and the adjacent basins.

The **numerical techniques** proposed take advantage of methods allowing non-structured meshes as finite volume and finite element methods. They appear to be suitable for the spatial discretization of problems that need to be solved in geometrically complex domains. They also permit the use of anisotropic mesh adaptation techniques. These techniques allow the automatic generation of meshes well adapted to the flow characteristics without dramatically increasing the computational cost. The use of adapted meshes makes it possible to capture phenomena along a wider spatial scale range by increasing the density of discretization points only on certain regions of the computational domain (see [11]).

This work has been undertaken in collaboration with the Universities of Santiago de Compostela and Seville (Spain), the Department of Applied Physics of the University of Malaga and the Instituto Español de Oceanografía.

The possible **applications** of this and other modelling and numerical simulation efforts may be:

- a better understanding of the hydrodynamical processes,
- helping in navigation,
- operational forecasting: marine accidents and contaminant spills, etc.,
- understanding the marine ecosystem,
- applications in Civil Engineering: pipelines, high tension cables, etc.

In the next section we describe some of the most important features that characterize the large scale dynamics in the Alboran Sea. The understanding of the physical problem and the main phenomena that must be reproduced by a model is the first necessary step in any modelling effort. This is extremely necessary in order to know which are the requirements the numerical model must possess to be able to represent these basic features. Section 3 is devoted to obtaining the numerical model and we briefly describe how the model is numerically solved. This model has been obtained by generalizing to multi-layer systems the shallow-water solver introduced by *Bermúdez et al. (1991)*, based on a mixed finite element method. To our knowledge, this is the first time a multi-layer finite element model has been used to study the water flow in the Alboran Sea. Besides this generalization, we have improved, in several ways, the performance of the shallow water solver in order to reduce the computational cost (see [13], [26], [31] for further details). In section 4 we show some numerical results. These results are briefly compared, from a qualitative point of view, with the main characteristics of the observed large scale dynamics described in section 2. Last section concludes with some final remarks.

2   The oceanographical problem

The **Alboran Sea** is, by its dynamics and by its economical and ecological importance, a very interesting area of the marine ecosystem. Being the western most part of the Mediterranean, it is the first basin to receive the Atlantic ocean water coming from the **Strait of Gibraltar**. It is also the last basin from which flows the Mediterranean water leaving this "sea between earths", feeding the deep current of the Strait. The Alboran Sea is, therefore, a transition basin characterized by interesting and specific dynamics.

The Mediterranean Sea is subjected to a particularly dry continental climate, even in wintertime, which causes intense evaporation throughout the year (*Tchernia, 1978*). The losses of water by evaporation exceed the gains due to precipitation and river contributions. This balance translates into an annual loss between 0.5 and 1 meters of water in the whole of the Mediterranean (*Béthoux, 1979; Harzallah, 1990*). Nevertheless, the characteristics of the Mediterranean waters do not appear to have changed over the previous centuries. Is the connection of the Mediterranean Sea with the Atlantic Ocean by the Strait of Gibraltar which produces an exchange that counteracts this deficit of water and thereby equilibrates the salt balance (*Lacombe et al., 1981*).

The mean exchange through the Strait is constituted by a superposed two-layer fluid. At the surface, Atlantic-origin less dense water flows eastwards, while, at depth, a much denser Mediterranean water penetrates into the Atlantic Ocean (*Lacombe and Richez, 1982*). This circulation satisfies a balance (Knudsen's relations) which ensures that the amount of water and salt in the Mediterranean remain constant (see *Bryden and Stommel, 1984* for further details). *Marsigli (1681)* showed, by performing an ingenious laboratory study of the interaction between two masses of water of different densities initially separated by a wall, that the water could return underneath as a dense underflowing current. Two centuries later, *Carpenter and Jeffreys (1870)* confirmed this hypothesis when they submerged a drag line 450 m down in the Strait of Gibraltar, and observed that the drag drifted to the west, in the opposite direction to the surface current.

Now it is common knowledge that the surface flux in the Strait is a consequence of the increase in density of the water masses produced in the Mediterranean basin under the effect of the ocean-atmosphere interactions. In effect, the Mediterranean Sea is a concentration basin that transforms the inflowing Atlantic water through the Strait into Mediterranean water (*Lacombe and Tchernia, 1972*). The Atlantic water, that is characterized by temperatures around 16°C and salinities about 36.5[1], penetrates in the Alboran Sea where it flows forming two large anticyclonic gyres (*Lanoix, 1974; Cheney and Doblar, 1982; Gascard and Richez, 1985; Kinder and Parrilla, 1991*). Afterwards, it continues along the Algerian coast (*Millot, 1985; Millot, 1991*). **Figure 1** schematizes the surface circulation described. A portion of these waters remains in the western Mediterranean where they accomplish a large cyclonic circuit, forming to the north the Liguro-Provenzal-Catalan current. Another portion of this water

---

[1] salinities will be expressed in the practical salinity scale (pss).

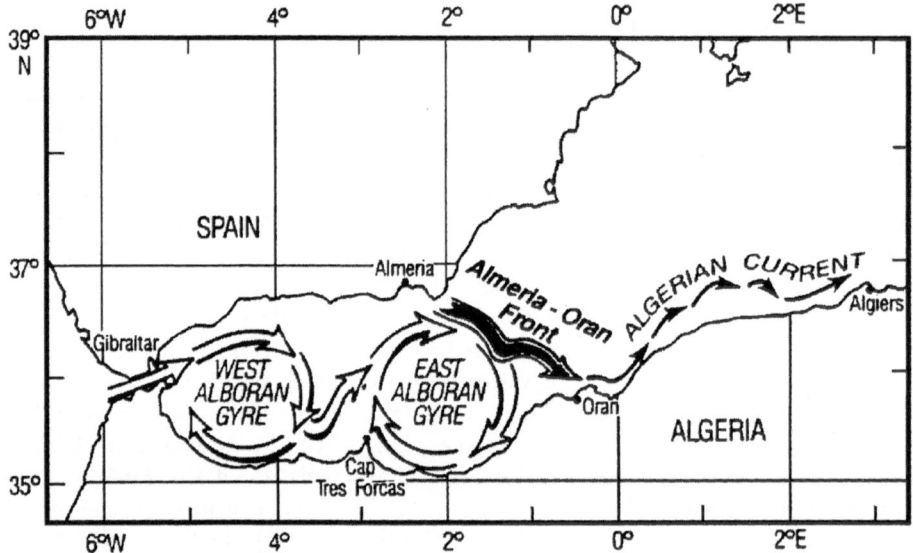

Fig. 1. Characteristic scheme of the dynamical structure of the circulation in the Alboran Sea. (From *Arnone et al., 1987*.)

makes its way to the eastern basin. In these two basins, the Atlantic-origin water is subject to transformation mechanisms by convective mixing, in response to winter atmospherical forcing.

Reciprocally, it is also well known the deep influence Mediterranean water has in the general circulation of the Atlantic Ocean and, in particular, in the formation of the North Atlantic Deep Waters (*Reid, 1979*). The Mediterranean water that penetrates in the Northern Atlantic Ocean creates a maximum in salinity at around 1000 meters depth. This saline contribution helps make the Atlantic Ocean water saltier than the waters of the Pacific and Indian Oceans.

Therefore, the Strait of Gibraltar plays an important role in the dynamics of the two basins that it connects and, especially, in the dynamics of the Alboran Sea. This sea is especially interesting as transition basin between the dynamics in the Strait and those of the rest of the Mediterranean.

2.1   The Alboran Sea: a transition basin

Numerous field studies (*Lanoix, 1974; Cano, 1977; Heburn and La Violette, 1990; Tintoré et al., 1991; Viúdez et al., 1996; ...*) show a quasi permanent circulation pattern in the western basin of the Alboran Sea dominated by the presence of an anticyclonic gyre approximately centered at 4° 10' west longitude, 35° 50' north latitude. This gyre has also been reproduced in both laboratory models (*Whitehead and Miller, 1979; Gleizon, 1994*), and numerical models (*Preller, 1986; Werner et al., 1988; Speich, 1992* among others).

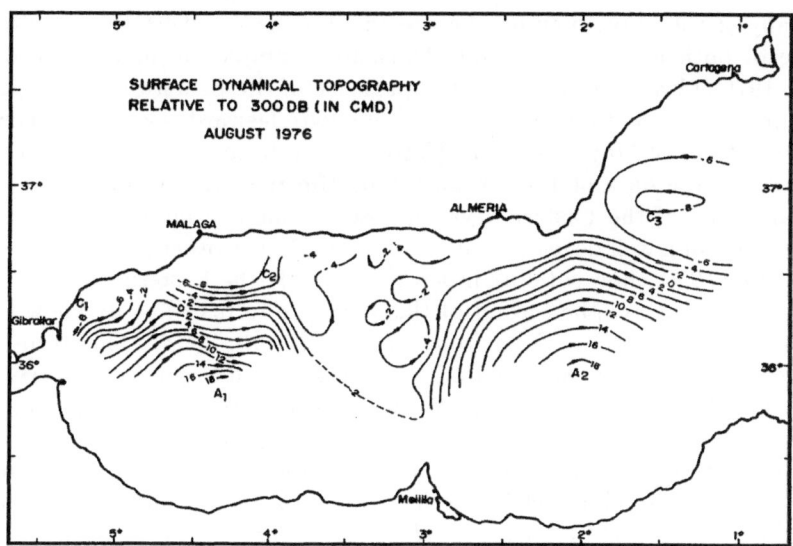

Fig. 2. Geostrophic currents obtained during the oceanographic cruise carried out by the O. V. Cornide de Saavedra in August 1976. From *Cano (1978a)*.

The main characteristics of the surface circulation in the Alboran Sea are depicted in **figure 2** representing the geostrophic circulation in the surface layer of the Alboran Sea obtained from data of a campaign of the IEO (Spanish Institute for Oceanography) in August 1976 (*Cano, 1978b*). The Atlantic jet feeds two large anti-cyclones that occupy, respectively, the western and eastern part of the Alboran Sea. The Atlantic water enters the Alboran Sea as a very intense surface current after having suffered an acceleration and a reduction in its thickness at its exit from the Strait (*Farmer and Armi, 1988; Perkins et al., 1990*). This Atlantic jet constitutes the northern border of the Western Gyre, creating a saline front at its entrance into the Alboran Sea. A portion of the Atlantic water that penetrates through the Strait forms, in most of the observations, a first anticyclonic gyre located just to the south of the incoming jet. This jet may extend up to the African coasts (*Kinder and Parrilla, 1997*). The rest of this mass of water follows the Moroccan coast, moving to the east as a coastal current and engendering, sometimes, a large anticyclonic meander before continuing its way along the African coast in the Algerian basin (*La Violette, 1985; Tintoré et al., 1988; Heburn and La Violette, 1990*).

A schematic picture of the dynamics described can be seen in **figure 1**. Nevertheless, a remarkable difference between the two gyres represented in figure 1 has to be noted: while the Western Gyre is quasi-permanent, the Eastern Gyre possesses a higher variability. *Herburn and La Violette (1990)*, from satellite imagery, statistically study the presence of the two anticyclonic gyres of the Alboran Sea, concluding that both gyres can disappear. These authors find situations where any of the gyres may not be present, but never where the simultaneous absence of both gyres were found. Thus, for example, an analysis of data of August 1976 (*Cano, 1978b*) shows the presence of the Eastern Gyre (**figure 2**), this appears reduced to a small anticyclonic gyre leewards Cape Tres Forcas in the observations of May–June 1973 (**figure 4(a)**, from *Cano, 1977*) and it even disappears in the data of July–August 1962 (**figure 4(b)**, from *Lanoix, 1974*). This variability of the Eastern Gyre opposes to the continuous presence, in all these studies, of a well developed Western Gyre. In the other hand, *Herburn and La Violette (1990)* show satellite images where only the Eastern Gyre is present. The eastern branch of the second of this two gyre forms what its known in the literature as the Almería–Oran front (*Tintoré et al., 1988*). This front, when it is present, defines an eastern limit for the Alboran basin (see **figure 1**). *In situ* studies during the campaigns of the "Western Mediterranean Circulation Experiment" (WMCE) showed that the Almería–Oran front was characterized by a strong density gradient, confined to the upper 300 m and was associated with an intense baroclinic jet in the first 50 to 75 meters with surface currents of the order of $0.6\,\mathrm{m\,s}^{-1}$ (*Arnone et al., 1987; Tintoré et al., 1988*).

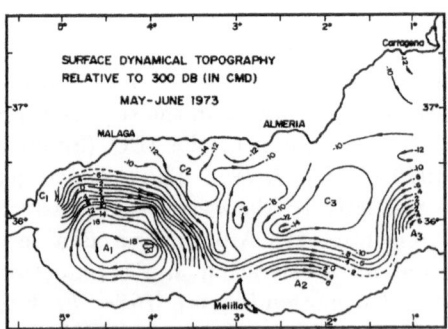 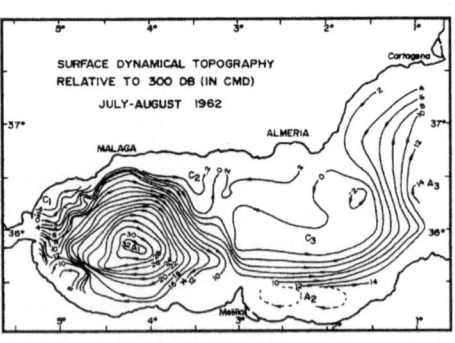

(a) In May–June 1973. From *Cano (1977)*.

(b) In July–August 1962. From *Cano (1978a)*.

Fig. 3. Geostrophic currents obtained during the oceanographical campaign by the O. V. Cornide de Saavedra.

3   The numerical model

3.1   Primitive equations

In Oceanography, it is usually accepted that the circulation of water mass is governed by the following system of P.D.E.:

$$\partial_t \boldsymbol{u}_h + [\boldsymbol{u} \cdot \nabla \boldsymbol{u}]_h + \frac{1}{\rho_0} \nabla_h p + f \boldsymbol{k} \times \boldsymbol{u}_h - F^{\boldsymbol{u}}(\boldsymbol{u}) = 0, \tag{1}$$

$$\partial_z p = -\rho g, \tag{2}$$

$$\nabla \cdot \boldsymbol{u} = 0, \tag{3}$$

$$\partial_t T + \nabla \cdot (T\boldsymbol{u}) = F^T(T), \tag{4}$$

$$\partial_t S + \nabla \cdot (S\boldsymbol{u}) = F^S(S), \tag{5}$$

$$\rho = \rho(T, S, p). \tag{6}$$

In this system the unknowns are: $\boldsymbol{u}_h$, the horizontal component of the three-dimensional velocity $\boldsymbol{u}$; the temperature, T; the salinity, S; the vertical component of the velocity, w; the density, ρ and pressure, p. Other notations in these equations are: $\boldsymbol{k}$ the local vertical axis, f the Coriolis parameter, g the acceleration due to gravity, ρ_0 a reference value for the density and $F^{\boldsymbol{u}}$, F^T and F^S that represent parameterizations of the effects of dissipation due to molecular viscosity or sub-grid mixing processes.

These equations, the so-called **primitive equations**, are derived from the incompressible **Navier–Stokes equations** using the **Boussinesq approximation**, in which density variations are neglected everywhere but in the gravity term. Density is related to the temperature, salinity and pressure through an equation of state (6).

Another hypothesis made in Oceanography is the **hydrostatic approximation**, which considers fluid vertical acceleration negligible compared with gravity-buoyancy effects. Therefore, vertical pressure gradient equilibrates with Boussinesq force. This is represented by equation (2). This hypothesis eliminates convective processes from the primary Navier–Stokes equations, which means that in a three-dimensional model they must be parameterized, and in a 2D model these processes are not represented. Besides, using this approximation allows only long waves (the so-called shallow water waves) to be simulated.

Equation (3) represents the incompressibility hypothesis (the three-dimensional divergence of the velocity vector is assumed to be zero), and the fourth and fifth equations are convection-diffusion equations for the temperature and salinity. The domain where the former system of P.D.E. has to be solved is defined by

$$\mathcal{O}_t = \Big\{ (x_1, x_2, x_3) \mid (x_1, x_2) \in \Omega, \; b(x_1, x_2) < x_3 < s(x_1, x_2, t) \Big\},$$

that represents the three-dimensional region between the sea bottom and the free sea surface.

In order to have a complete problem, some initial and boundary conditions must be considered depending on the particular problem. Remark that the atmospheric effects input the ocean model as boundary conditions through the sea surface.

The **difficulties** encountered in these equations, both from the mathematical and the numerical point of view, are manifold: non linearities, geometrically complex three-dimensional domains, free surfaces, turbulent regimes, boundary layers, anisotropy between horizontal and vertical movements or the wide range of space-time scales involved. Due to all these difficulties, it is a common practice in Oceanography (and other branches of Science) to turn to **simplified models**.

A usual simplification made in Oceanography and Climatology is the so-called **shallow water approximation**, that consists in obtaining from the primitive equations a bidimensional system by means of a procedure of vertical integration of the equations. This is the approximation that we have followed in the present study. More precisely, we suppose that the water column is composed of several inmiscible layers of fluid with different constant densities and consider a shallow-water approximation at each of these layers of water. Therefore, a multi-layer model is deduced. For this approximation to be valid, wavelengths of the phenomena to be simulated must be, roughly speaking, much larger than the thickness of the water layer.

In the next section, we introduce the one-layer model. Then the model is generalized to a multi-layer configuration and used in its two-layer version to simulate the large-scale dynamics in the Alboran Sea.

3.2   The one-layer model equations

Before introducing model equations, let us consider the notations given in the following figure.

$h(x_1, x_2, t) = s(x_1, x_2, t) - b(x_1, x_2)$ represents the thickness of the water layer, η is the elevation of the free sea surface above a chosen reference level A (for example the mean sea surface height) and H is the bathymetry of the basin (depth from the reference level to the sea bottom), if the reference level has been chosen to be the mean sea surface.

As the three-dimensional primitive equations must be vertically integrated to deduce the two-dimensional shallow-water model, we introduce the mean velocity vertically integrated $\bar{u}$, defined as:

$$\bar{u}(x_1, x_2, t) = \frac{1}{h} \int_b^{b+h} u_h(x_1, x_2, x_3, t)\, dx_3.$$

After the vertical integration process and some further simplification hypotheses (see [3], [12], [25] for details on the deduction of model equations), the system of equations obtained can be written in conservative form ($Q = h\bar{u}$) as:

$$\frac{\partial \eta}{\partial t} + \nabla \cdot Q = 0 \quad \text{in } \Omega, \tag{7}$$

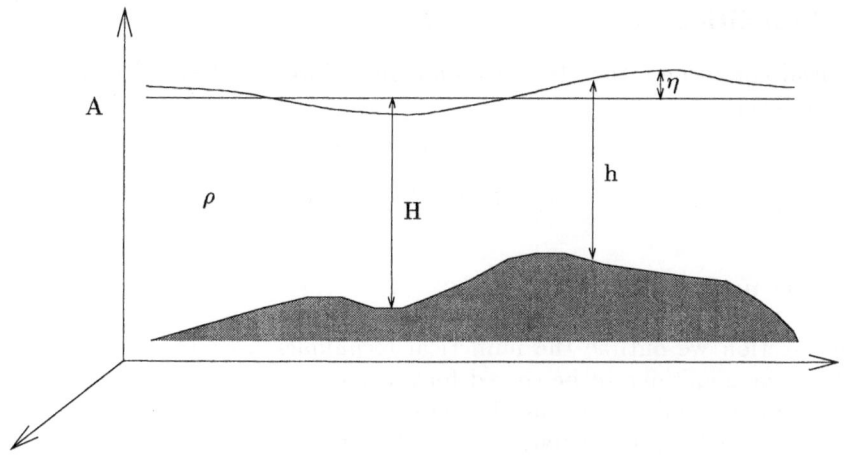

Fig. 4. Notations for the one-layer model

$$\frac{\partial Q}{\partial t} + \nabla \cdot \left[(\bar{u} \otimes Q) + \frac{1}{2}g(\eta^2 + 2\eta H)\delta \right] = g\eta\nabla H + F + \frac{1}{\rho_0}(\tau_w - \tau_f) \quad \text{in } \Omega, \quad (8)$$

where Ω represents the horizontal projection of the three-dimensional domain $\mathcal{O}_t$, δ is the unit tensor, $F = f(Q_2 - Q_1)$ the term due to Coriolis effects, $\tau_w = \gamma|v|v$ is the term parameterizing wind stress effects and $\tau_b = \dfrac{g|\bar{u}|}{C^2}\bar{u}$ parameterizes bottom drag effects.

This model was originally used by *Bermúdez, Rodríguez, Vilar (1991)* to simulate tidal effects in the Galician Rias.

Boundary conditions

The boundary conditions permitted by the model are:

(a) Normal component of the flux given,

$$Q \cdot n = f,$$

where n is the normal outward unit vector to the boundary of Ω, $\partial\Omega$. This condition is useful on the inflow/outflow boundaries when the flux is known, and on coastal boundaries where $f = 0$.

(b) Elevation given,

$$\eta = \varphi.$$

It is used on the inflow/outflow boundaries, when the elevation is known. This condition is useful, for instance, to simulate the effects due to tide fluctuations.

Initial conditions

As initial conditions, the elevation and mean integrated velocity initial state must be given:

$$\eta(x, 0) = \eta_0(x) \quad \text{in } \Omega,$$

$$\bar{u}(x, 0) = \bar{u}_0(x) \quad \text{in } \Omega.$$

3.3   The numerical scheme

In this section we outline the numerical techniques used to numerically solve (7)–(8). The equations to be solved form a bidimensional system of non-linear first-order hyperbolic equations. Therefore, their properties are similar to those of the compressible Euler equations. As it is well know, schemes based on explicit Euler discretization in time together with centered differences for flux terms are unconditionally unstable for hyperbolic equations even in the linear case. The upwind discretization should be used. In [3] a method combining the characteristics method to discretize convective terms with first-order Raviart–Thomas finite elements for the space discretization is given. This is the method we have used and extended to multi-layer systems.

3.3.1 Time discretization

The mass conservation equation (7) is discretized using an **Euler implicit** method, obtaining the discretized equation,

$$\eta^{n+1} = \eta^n - \Delta t\, \nabla \cdot \mathcal{Q}^{n+1}$$

where we use the standard notations: Δt is the time step and η^n, $\mathcal{Q}^n$ are the approximations of the elevation η and the flux $\mathcal{Q}$ at time $n\Delta t$.

To discretize the convective term in the momentum equation (8) the **method of characteristics** (cf. [33]) is used. This method is based on the fact that the first two addends in the momentum equation (8) coincide with the material derivative of the product $J\mathcal{Q}$, i.e.,

$$\frac{D}{Dt}(J\mathcal{Q}) = \frac{\partial \mathcal{Q}}{\partial t} + \nabla \cdot (\bar{u} \otimes \mathcal{Q}),$$

where $\frac{D}{Dt} = \frac{\partial}{\partial t} + u \cdot \nabla$ is the **material** or **convective derivative**, i.e., the derivative following particle trajectories and J represents the evolution of the volume element, i.e., is the Jacobian of the transformation associated to the flux, that is characterized as being the solution of the differential equation

$$\begin{cases} \dfrac{d}{d\tau} J(x, t; \tau) = \nabla \cdot u\left(X(x, t; \tau), \tau\right) J(x, t; \tau) \\ J(x, t; t) = 1. \end{cases}$$

Here the function $\tau \to X(x,t;\tau)$ represents the trajectory of the particle that is located at a point x at time t, and therefore $X(x,t;\tau)$ is the solution of

$$\begin{cases} \dfrac{d}{d\tau}X(x,t;\tau) = u\left(X(x,t;\tau),\tau\right) \\[2mm] X(x,t;t) = x. \end{cases}$$

The idea behind the method consists of applying a scheme of type backward finite differences following the trajectories. Observe that while at time $n+1$ functions are evaluated at a point x, at the previous time step n they are evaluated at $X^n(x)$, which represents the position occupied at time n by a particle located at the point x at time $n+1$. This method has good properties of stability and eliminates the non-linearity of the convective term.

3.3.2 Discretized equations

Using the two time discretization schemes described above, the semi-discretized equations are written as follows:

Given Q^n, η^n :

$$\frac{Q^{n+1}}{\Delta t} + \frac{g}{2}\nabla\left((\eta^{n+1})^2 + 2\eta^{n+1}H\right) - g\eta^{n+1}\nabla H = \frac{J^n Q^n [X^n]}{\Delta t},$$

$$\eta^{n+1} = \eta^n - \Delta t\,\nabla\cdot Q^{n+1} \quad \text{in } \Omega.$$

In order to simplify the notation (dropping the n's), the problem to be solved at each time iteration can be rewritten as

$$\frac{Q}{\Delta t} + \frac{g}{2}\nabla\left(\eta^2 + 2\eta H\right) - g\eta\nabla H = F \quad \text{in } \Omega,$$

$$\eta = \eta_0 - \Delta t\,\nabla\cdot Q \quad \text{in } \Omega,$$

$$Q\cdot n = 0 \quad \text{at } \partial\Omega,$$

where η_0 denotes the elevation at the previous time step and F represents the term given by the method of characteristics.

Although the method of characteristics eliminates the non-linearity of the convective term, the non-linearity due to the pressure term remains. In fact the **difficulty** for numerically solving this system of equations comes from this non-linearity $(\eta^2 + 2\eta H)$, that is a maximal-monotonous operator, which allows the use of a numerical algorithm proposed by *Bermúdez and Moreno (1981)* (see [3] or [25] for further details on the numerical resolution of model equations).

The space discretization is performed using the first-order Raviart–Thomas Finite Element. The degrees of freedom for the flux Q are the values on the middle points of the edges of the elements and the elevation η is constant by triangle. The resulting linear systems (with non-symmetric matrices) are solved by a Stabilized Conjugate Bigradient preconditioned with an LU incomplete type factorization (see [25] and [13] for further details).

The main drawbacks of the chosen algorithm derive from the implicit nature of the discretization performed, and from the restrictions on the type of boundary conditions that can be imposed. On the one hand, it is known that implicit schemes introduce additional damping effects on the numerical solution and increase the computational cost when compared with explicit schemes. The advantage of the chosen discretization is that it allows dealing with the appearance of regions where the thickness of the water layer vanishes (see [25], [26]). Moreover, at this first stage, we only look for obtaining steady states or solutions without sharp gradient regions, and damping effects are assumed not to be important. Nevertheless, this algorithm provides good results in more general cases if small enough time steps are chosen. Therefore, in order to have an efficient algorithm, the computational cost of solving the non-linear problems appearing at each time step must be small. In order to have this, we have accelerated the original algorithm by implementing an automatic choice of parameters (see [25] or [31]). Concerning the boundary conditions, the algorithm used to solve the non-linear problems include an elliptic regularization of the problem. Due to this regularisation, only one condition can be imposed on each part of the boundary. Therefore, supercritical flows close to the input/output boundaries cannot be handled. Currently we are working on the adaptation of the algorithm to these conditions.

3.4   The multilayer model

Once the one-layer model has been introduced, we undertake the description of the multilayer model. For that case we consider the notations given in the following figure representing, for the sake of simplicity, a two-layer configuration:

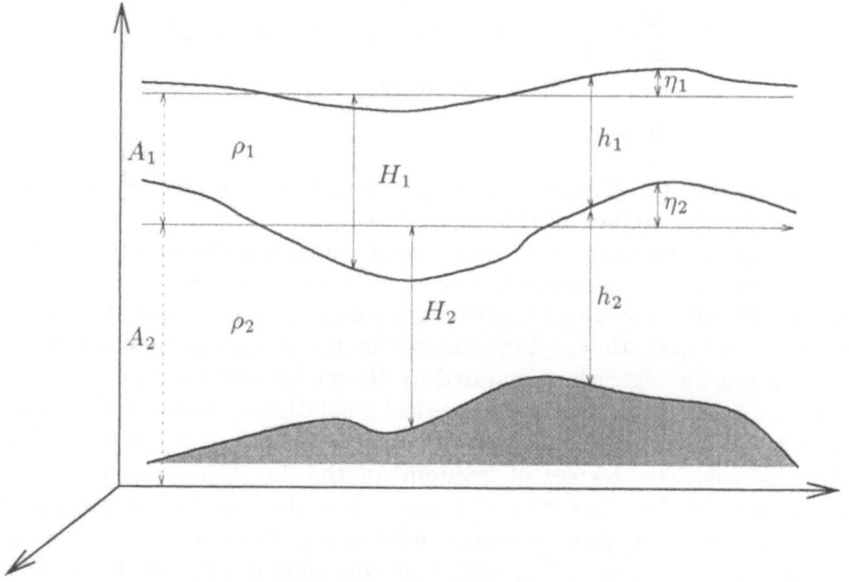

Two-layer model scheme and notations

As it was pointed out in the section devoted to the introduction of the physical problem, a one-layer model cannot represent the observed dynamics in the Alboran Sea and the Strait of Gibraltar (basically composed by a two-layer fluid with different densities flowing in opposite directions). Therefore, a one-layer model is not admissible in that case. The multilayer model will suppose the water column composed of several layers of water. At each layer, a shallow water approximation is considered, i.e.: the primitive equations are written at each layer with suitable boundary conditions at the interface and are vertically integrated at each of the layers. Doing so, a coupled shallow water system is obtained where the coupling takes place through the pressure and friction terms. In the formulation used in this study, this system of equations writes as follows:

First layer

$$\frac{\partial h_1}{\partial t} + \nabla \cdot Q_1 = 0,$$

$$\frac{\partial Q_1}{\partial t} + \nabla \cdot \left[(\bar{u}_1 \otimes Q_1) + \frac{1}{2} g (\eta_1{}^2 + 2\eta_1 H_1) \delta \right] = g\eta_1 \nabla H_1 + F_1 + \frac{1}{\rho} (\tau_w - \tau_{i_1}).$$

k^{th}-layer

$$\frac{\partial h_k}{\partial t} + \nabla \cdot Q_k = 0,$$

$$\frac{\partial Q_k}{\partial t} + \nabla \cdot \left[(\bar{u}_k \otimes Q_k) + \frac{1}{2} g (\bar{\eta}_k^2 + 2\bar{\eta}_k \bar{H}_k) \delta \right] = g\bar{\eta}_k \nabla \bar{H}_k + F_k + \frac{1}{\rho} (\tau_{i_{k-1}} - \tau_{i_k}),'$$

- $\bar{\eta}_k = \eta_k + \dfrac{\rho_1}{\rho_k} h_1 \cdots + \dfrac{\rho_{k-1}}{\rho_k} h_{k-1},$

- $\bar{H}_k = H_k - \dfrac{\rho_1}{\rho_k} h_1 - \cdots - \dfrac{\rho_{k-1}}{\rho_k} h_{k-1}.$

Lower layer

$$\frac{\partial h_n}{\partial t} + \nabla \cdot Q_n = 0,$$

$$\frac{\partial Q_n}{\partial t} + \nabla \cdot \left[(\bar{u}_n \otimes Q_n) + \frac{1}{2} g (\bar{\eta}_n^2 + 2\bar{\eta}_n \bar{H}_n) \delta \right] = g\bar{\eta}_n \nabla \bar{H}_n + F_n + \frac{1}{\rho} (\tau_{i_{n-1}} - \tau_b),$$

- $\bar{\eta}_n = \eta_n + \dfrac{\rho_1}{\rho_n} h_1 + \cdots + \dfrac{\rho_{n-1}}{\rho_n} h_{n-1},$

- $\bar{H}_n = H_n - \dfrac{\rho_1}{\rho_n} h_1 - \cdots - \dfrac{\rho_{n-1}}{\rho_n} h_{n-1}.$

The new variables $\bar{H}$ and $\bar{\eta}$ are introduced in order to formally obtain the same equations at each layer and, therefore, to enable us to apply the same algorithm of resolution in all the layers. The unknowns to be computed in this system of equations are Q_k, the fluxes at each layer, and η_k, the elevations at the different interfaces.

In practice, we will apply the **two–layer model** for simulating the dynamics in the Alboran Sea. In the next section some numerical results are presented.

4   Some numerical results

In this section we depict some figures to illustrate model results. The experiments performed were aimed to study the two Alboran Sea gyres and their variations to different wind conditions. A complete set of figures and comments can be found in [25], here we restrict ourselves to a single numerical example where no wind conditions were imposed.

In the experiment shown here the two layer model have been used. The upper layer (initially at 80 m depth) represents the inflowing Atlantic water that enters through the Strait of Gibraltar and exits into the western Mediterranean basin. In this layer the constant value for the water density was taken to be equal to 1027 kg m^{-3}. The lower layer represents the Mediterranean water pouring from the eastern Mediterranean into the Atlantic. The constant density value taken in this layer was 1029 kg m^{-3}.

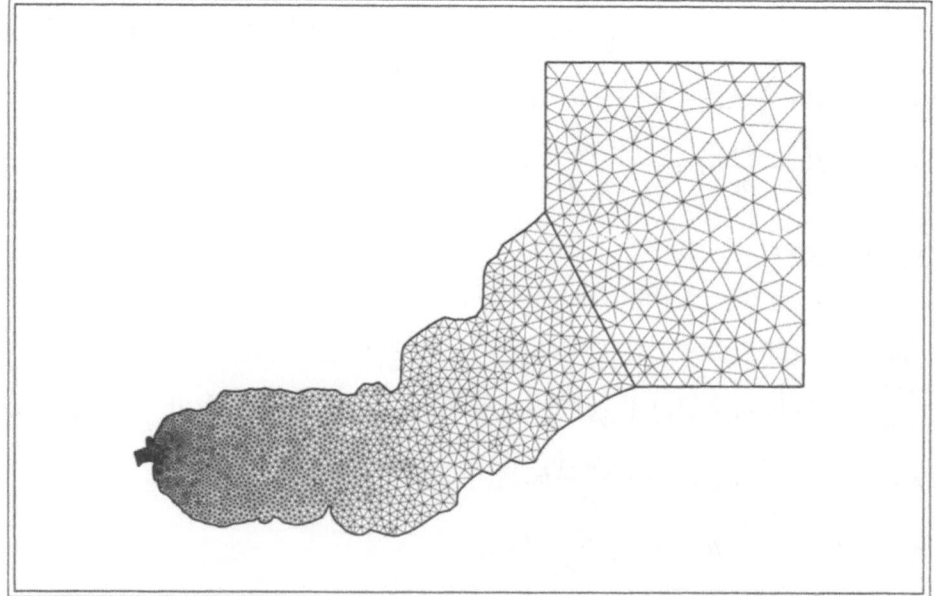

Fig. 5. Spatial domain considered and its finite element discretization. The mesh contains 4,792 triangles and 7,317 nodes.

The physical domain considered presented 4 different boundaries. Two "natural" boundaries, corresponding to the Spanish and Moroccan coasts and two "artificial" boundaries limiting the computational domain on the east and west. The western boundary near Tarifa, in the Strait of Gibraltar, and the eastern limit consisted of the sides of a large rectangle. These sides followed the orientation of the Spanish and African coasts at that part of the Mediterranean (see **figure 5**). The meshes were generated from digitalized cartographic data provided by the I.E.O. (Instituto Español de Oceanografía), using HYPACK code. When real bottom bathymetry was considered, the bathymetry function H was computed from digital cartographic data by means of an automatic interpolation process over the mesh vertices. From this discrete function and the corresponding first layer mesh, the second, and subsequent, layer meshes were automatically constructed suppressing the spare triangles, i.e., the elements, k in the first layer, with $H(k) > -80\,$m were suppressed. **Figure 6** depicts the second layer mesh obtained by this procedure from the first layer triangularization shown in fig. 5.

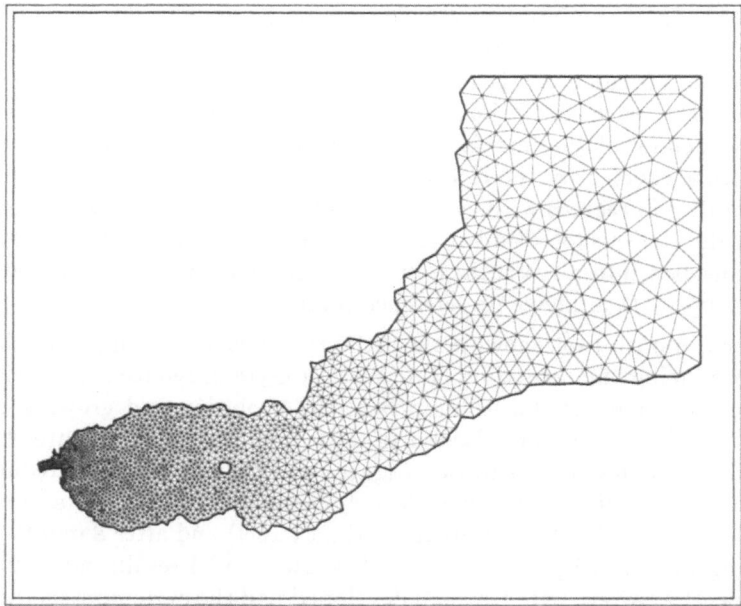

Fig. 6. Second layer finite element discretization, when real bottom bathymetry is taken into account. Otherwise, when constant depth is considered the second layer spatial discretization is the same as for the first layer. The mesh contains 4,094 triangles and 6,283 nodes.

As initial conditions, elevation and fluxes equal to zero have been imposed, i.e., start from a resting sea. At the first layer, the input flux was imposed over the Strait of Gibraltar. A total flux of $1\,\mathrm{Sv}$ ($1\,\mathrm{Sv} = 10^6\,\mathrm{m}^3\mathrm{s}^{-1}$) was taken, which is the estimate corresponding to the annual mean of the Atlantic input flux (see [5]). The profile considered for this input flow was designed to fulfil a criterion of conservation of the potential vorticity as described in [25]. On the sides of the rectangle, coastal conditions were imposed at both layers. For the second layer at the Strait, an output symmetric to the first layer input was imposed. To conserve mass at each layer it was necessary to consider an exchange between layers of 1 Sv. In the example presented here, this exchange was imposed to take place in the most eastern part of the domain, in the region limited by the rectangle. As external forcing, the wind can be imposed. In the example presented here no wind conditions are considered. Therefore, the energy needed for the system to move was exclusively provided by the input flow through the Strait of Gibraltar. Numerical experiments showing the effects of different wind conditions and starting from other initial conditions can be found in [25]. In the simulation shown here the time step is 15 min.

Figures 7 and 9 depict the time evolution of the first layer velocity field. Initially, an evolving "coastal mode" is developed. Figure 7 shows the velocity field after 10 (left panel) and 20 days (right panel) of integration. The elevations of the sea surface and interface reproduced by the model after 20 days of integration are shown in figure 8. It can be observed that, when an Atlantic jet flowing close to the African coast is simulated, an accumulation of water in this coast is produced by effect of Coriolis force. This reflects in a rising of the sea surface and a deepening of the interface (up to 37 m under the reference level) due to the pressure exerted by the thicker Atlantic layer. In the other hand, it can be observed that in regions with cyclonic circulation the free surface is depressed by effect of the Coriolis force and at the same time, the interface becomes shallower. This translates into a sea surface up to 13 cm below the mean sea level and an interface up to 15 m above its reference level.

After a month (not shown), the Atlantic jet is separating from the African coast, producing the formation of a Western Gyre of reduced dimensions, while at the eastern part of the Alboran Sea, the evolution and growing of another large anticyclonic structure also starts to be evident. This structure was already present after 10 days of integration (figure 7, left panel), although reduced to one half of its maximal size. Ten days latter, it has reduced its size and is confined leewards of Cape Tres Forcas (figure 7, right panel) and after a month it is again increasing in size. Figure 9 left panel, shows model results after 40 days. At that time, the eastern gyre was well developed and the cyclonic structure appears more confined to the north of the Atlantic jet. The eastern gyre was also formed. This figure and next one (figure 9, right panel) show a two-gyre configuration very similar to that depicted in figures 1 and 2 in the section devoted to the introduction of the physical problem.

Performing further in the time integration, the Western Gyre continues growing and the eastern one disappears. A one-gyre configuration is obtained, with the single gyre located to the west of Cape Tres Forcas (not shown). Figure 10 shows the elevations simulated after 50 days. It can be observed the clear sig-

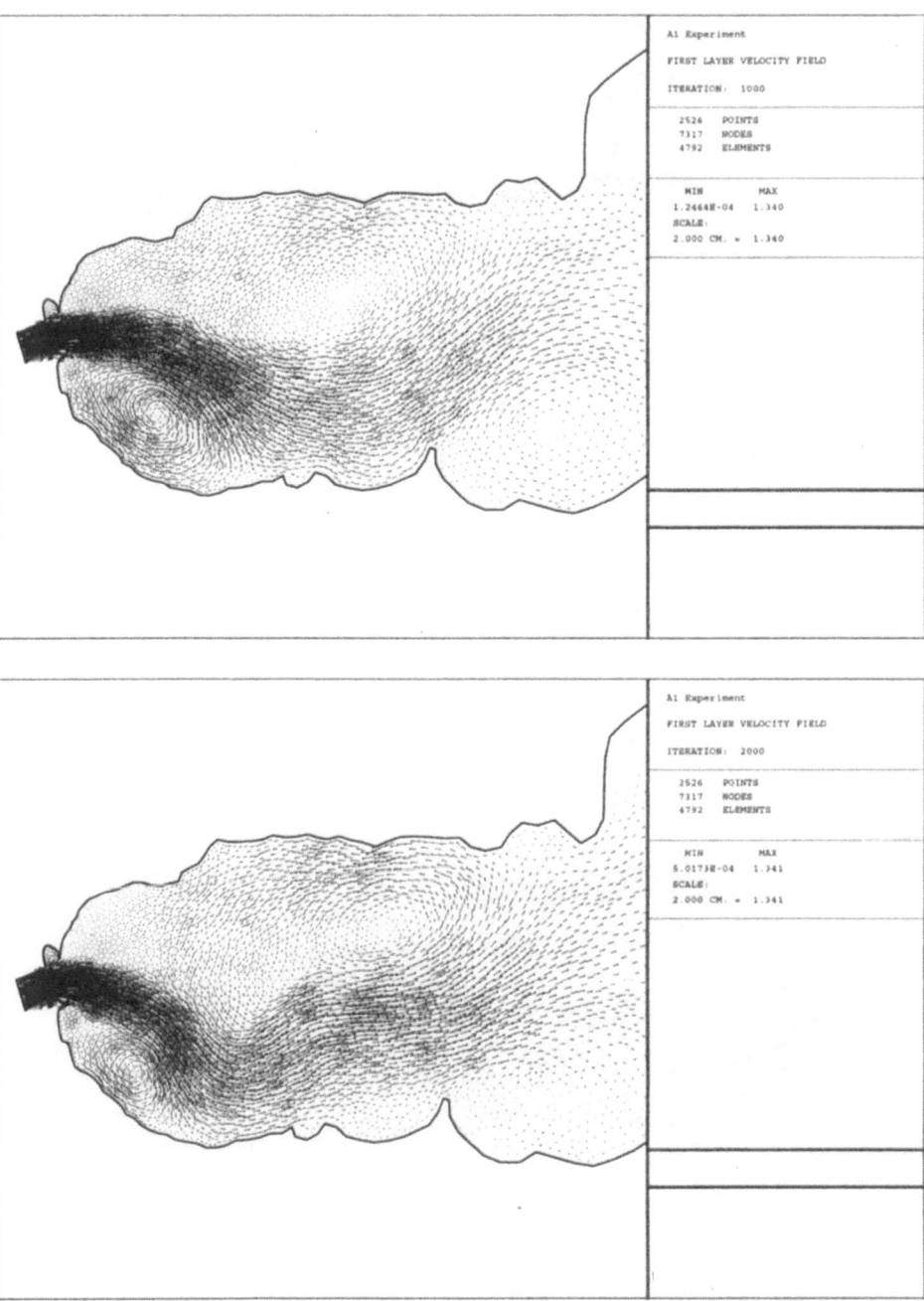

Fig. 7. First layer mean velocity field at time iterations 1,000 and 2,000.

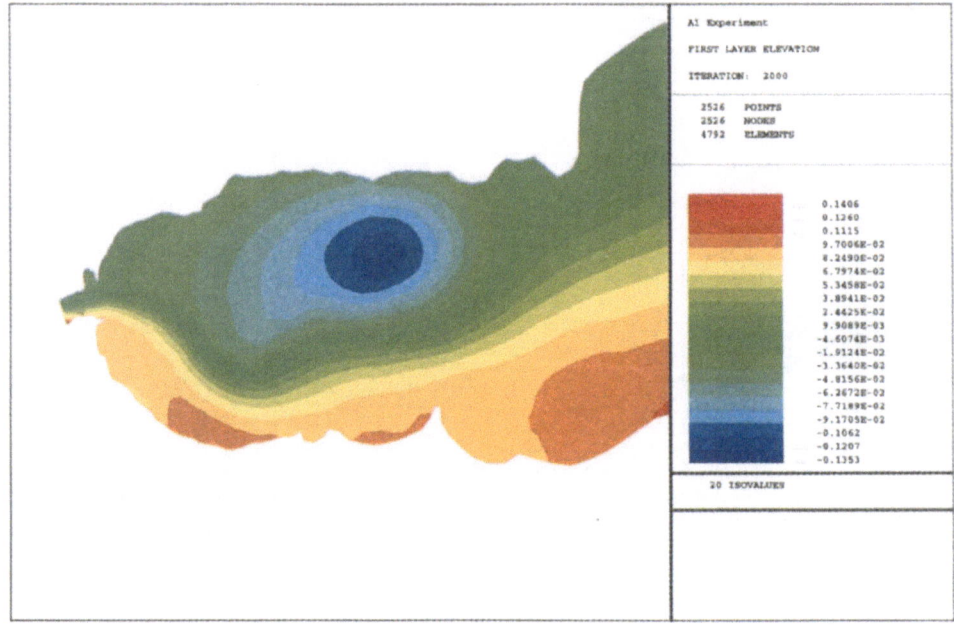

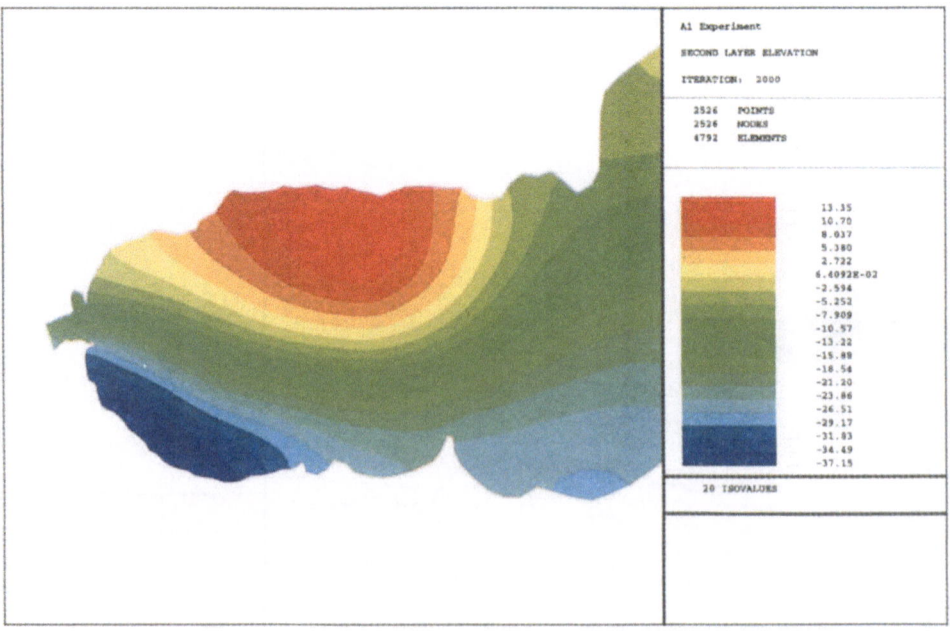

Fig. 8. Sea surface and interface elevation at iteration 2,000.

nature of the anticyclonic gyre on the sea surface elevation. On the other hand, the opposite effect is evident in the regions with cyclonic circulation.

The differences on the simulated sea level are of about 30 cm between higher and lower sea surface regions. Again it can be observed the effect of the rising/sinking of the sea surface on the location of the interface: below the anticyclonic gyres water accumulates and the interface sinks (up to 35 m below the reference level in this simulation); under the cyclonic gyres the interface is shallower (15 m above the reference level). It can be observed that the maximum sinking is not located exactly below the center of the anticyclonic gyres, as occurs with the sea surface elevations, but this is displaced southwards. This situation has also been reproduced by other models (see, for example, [36]) although in observational data this seems not to be the case. In what respect the sea surface, it would be interesting to dispose of suitable satellite data to compare and validate model results. Nevertheless, satellite data currently available are of very large spatial resolution to be used in regions of the dimensions of the Alboran Sea (reduced at global scale). In the other hand, for such a comparison it might be taken into account that the sea surface elevations simulated by the model correspond to a dynamical topography of the sea surface in which the effects of the atmospheric pressure are not included.

5 Final remarks

The development of the model presented fits into a wider project whose objective is the modelling and numerical simulation of the Alboran Sea and the Strait of Gibraltar dynamics at different spatial and temporal scales, by means of the use of various models, among them the shallow-water multi-layer model presented here. At an initial stage, the aim of this project was to find solutions that qualitatively approximate the large-scale structures characterizing the dynamics in the Alboran Sea. In view of the all numerical results obtained, some of which are outlined in the text, it seems that the model developed can be used as a tool to better understand the physical problem set by the Strait of Gibraltar and the Alboran Sea environment. Nevertheless, it must be pointed out that it seems suitable, for a deeper study of the dynamics simulated by the model, as well as to understand the observed dynamics, to perform a set of numerical experiments, which would include testing the effects of different winds, variable in time and/or space, or the effect of time variations in the input Atlantic flow, together with other characteristics of this flow, such as its vorticity.

In order to improve the model, we are carrying out the following research activities:

- Dealing with vanishing thickness regions and using penalisation techniques to improve convergence.
- Generalizing the choice of boundary conditions. Study the possibility of using data assimilation techniques.
- Implementation and comparison of another time and/or space discretization, including second-order schemes.

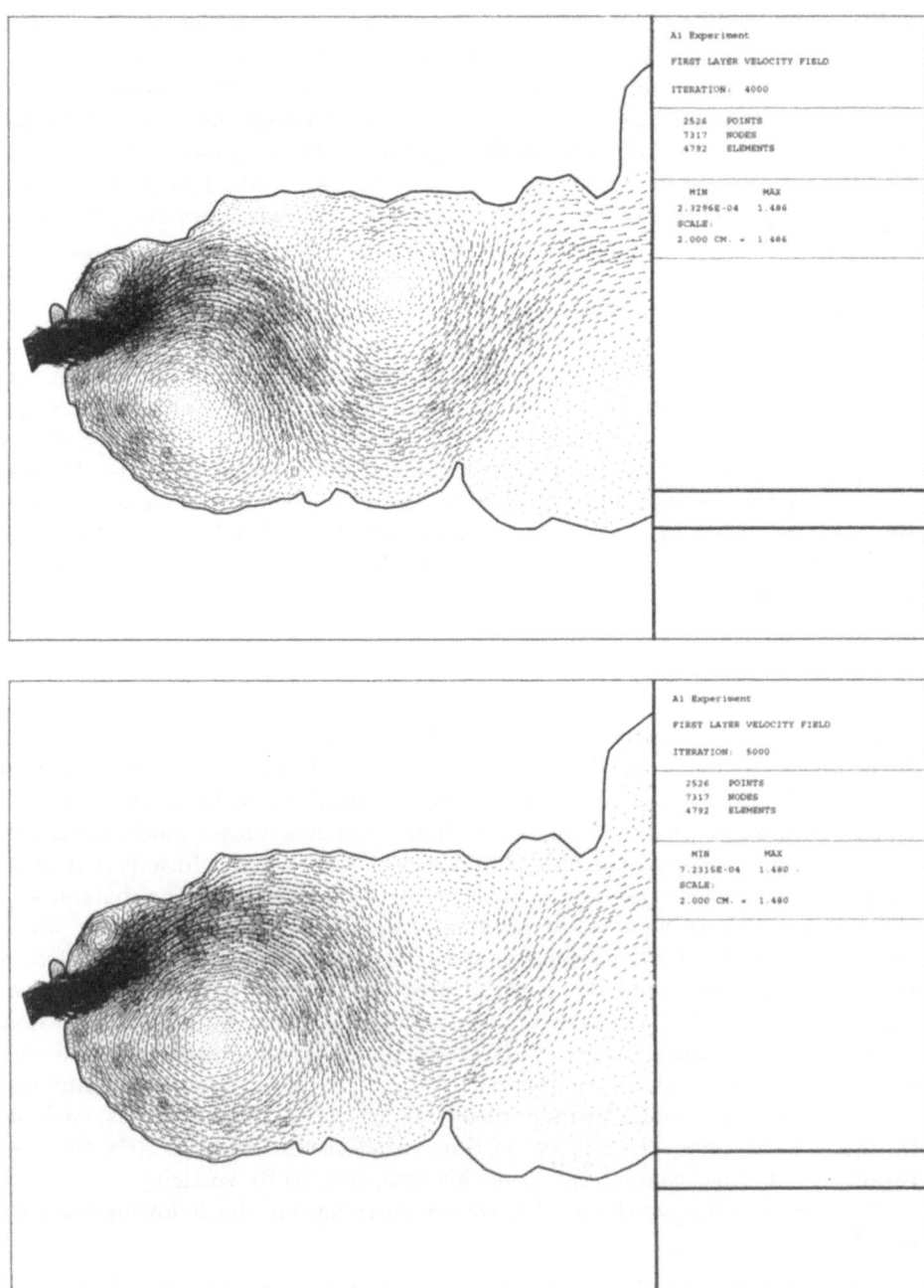

Fig. 9. First layer mean velocity field at time iterations 4,000 and 5,000.

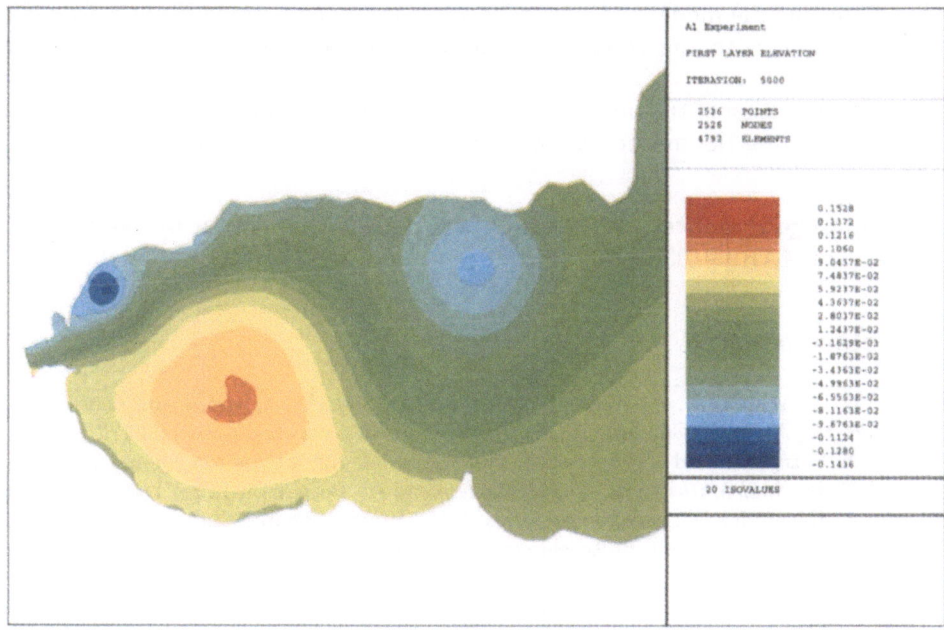

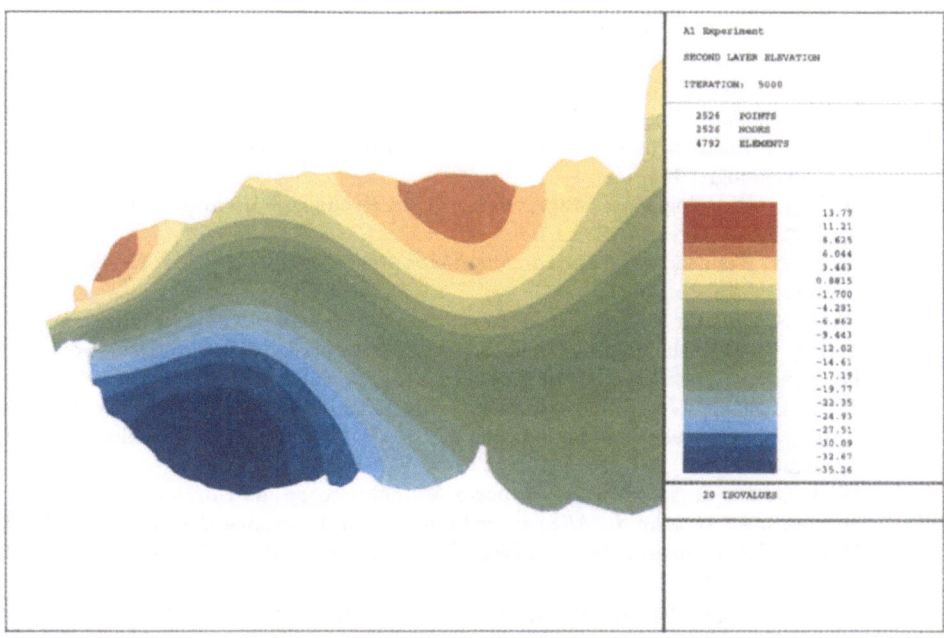

Fig. 10. Sea surface and interface elevation at iteration 5,000.

- Testing the model under more realistic wind and boundary conditions. Comparison with the results with previous works and experimental measurements.
- Application of the model to the study of the variability of the Alboran gyres.
- Analysing the mathematical model.

Acknowledgments: This research was partially supported by the C.I.C.Y.T. (project MAR97-1055-CO2-01).

References

1. R. A. Arnone, D. A. Wiesenburg, and K. D. Saunders. Origins and characteristics of the Algerian Current. *EOS, Trans. Americ.Geophys. Union*, 68:1725, 1987.
2. A. Bermúdez and C. Moreno. Duality methods for solving variational inequalities. *Comp. and Maths. with Appls.*, 7:43–58, 1981. Pergamon Press Ltd.
3. A. Bermúdez, C. Rodríguez, and M. A. Vilar. Solving shallow water equations by a mixed implicit finite element method. *IMA Journal of Numerical Analysis*, 11:79–97, 1991.
4. J. P. Béthoux. Budgets of the Mediterranean Sea: Their dependence on the local climate and on the characteristics of the Atlantic waters. *Oceanol. Acta*, 7(3):289–296, 1979.
5. H. Bryden, J. Candela, and T. H. Kinder. Exchange through the Strait of Gibraltar. *Prog. Oceanogr.*, 33:201–248, 1994.
6. H. L. Bryden and H. M. Stommel. Limiting processes that determine basic features of the circulation in the Mediterranean Sea. *Oceanol. Acta*, 7(3):289–296, 1984.
7. N. Cano. Resultados de la campaña "Alborán 73". Boletín 1 (230):103-177, Inst. Esp. Oceanogr., 1977.
8. N. Cano. Hidrología del Mar de Alborán en Primavera-Verano. Boletín 4(248):51-66, Inst. Esp. Oceanogr., 1978a.
9. N. Cano. Resultados de la campaña "Alborán 76". Boletín 4 (247):3-49, Inst. Esp. Oceanogr., 1978b.
10. W. B. Carpenter and J. G. Jeffreys. Report on deep-sea researches carried· on during the months of July, August, and September 1870 in HM surveying-ship Porcupine. *Proceedings of the Royal Society, London*, 19:146–221, 1870.
11. M. J. Castro. *Generación y Adaptación Anisótropa de Mallados de Elementos Finitos para la Resolución Numérica de E.D.P.* PhD thesis, Universidad de Málaga, Noviembre 1996. 216 pp.
12. M. J. Castro and J. Macías. *Modelo Matemático de las Corrientes Forzadas por el Viento en el Mar de Alborán*, volume 5. Publicaciones del Grupo de Análisis Matemático Aplicado de la Universidad de Málaga, 1994. 350pp, ISBN:84-7496-252-8.
13. M. J. Castro, J. Macías, and C. Parés. An incomplete LU-based family of preconditioners for numerical resolution of a shallow water system using a duality method. Applications. *Appl. Math. Lett.*, 14:651-656, 2001.
14. R. E. Cheney and R. A. Doblar. Structure and variability of the Alboran Sea frontal system. *J. Geophys. Res.*, 87(C1), 1982.
15. J. C. Gascard and C. Richez. Water masses and circulation in the western Alboran Sea and in the Strait of Gibraltar. *Progr. Oceanogr.*, 15:157–215, 1985.

16. P. Gleizon. *Étude Experimentel de la Formation et de l'Estabilité de des Tourbillons Anticycloniques Engendres par un Courant Barocline Issu d'un Detroit. Application a la Mer d'Alboran.* PhD thesis, Université Joseph Fourier, Grenoble I, 1994. 240 pp.

17. A. Harzallah. *Étude aérologique et océanique de l'hydrologie du bassin méditerranéen.* PhD thesis, Université Paris VI, Paris, 1990. 212pp.

18. G. W. Heburn and P. E. La Violette. Variations in the structure of the anticyclonic gyres found in the Alboran Sea. *J. of Geophys. Res.*, 95(C2):1599–1613, February 1990.

19. T. H. Kinder and G. Parrilla. The summer 1982 Alboran Sea gyre. *J. Geophys. Res.*, Accepted, 1997.

20. P. E. La Violette. Western Mediterranean Circulation Experiment Operation Plan. WMCE Newsletter 5, 1985. 48pp.

21. H. Lacombe, J.-C. Gascard, J. Gonella, and J. Béthoux. Response of the Mediterranean Sea to water and energy fluxes across its surface, on a seasonal and interannual scales. *Oceanol. Acta*, 4(2):247–255, 1981.

22. H. Lacombe and C. Richez. The regimen of the Strait of Gibraltar. In J. C. J. Nihoul, editor, *Hydrodynamics of Semi-Enclosed Seas*, pages 13–73. Elsevier, Amsterdam, 1982.

23. H. Lacombe and P. Tchernia. Caractères hydrologiques et circulation des eaux en Mediterranée. In D. J. Stanley, editor, *The Mediterranean Sea: A Natural Sedimentation Laboratory.* Dowden, Hutchinson and Ross, Stroudsburg, Pennsylvania, 1972. 765 pp.

24. F. Lanoix. Project Alboran. Étude hydrologique et dynamique de la Mer d'Alboran. Rapport Technique 66, OTAN, Brussels, 1974.

25. J. Macías. *Some Topics in Numerical Modelling in Oceanography.* PhD thesis, University of Paris VI, Paris, November 1998.

26. J. Macías, C. Parés, and M. J. Castro. Improvement and generalization of a shallow-water solver to multilayer systems. *Int. J. Numer. Methods Fluids*, 31:1037–1059, 1999.

27. L. F. Marsigli. Internal observation of the Thracian Bosphorus, or true channel of Constantinople, represented in letters to her majesty, Queen Christina of Sweden (Translated by E. Hudson). In M.B. Deacon, editor, *Oceanography: concepts and history*, 394 pp. Dowden, Hutchinson and Ross, Stroudsburg, Pensylvania, 1681.

28. C. Millot. Some features of the Algerian Current. *J. Geophys. Res.*, 90:7169–7176, 1985.

29. C. Millot. Mesoscale and seasonal variabilities of the circulation in the Western Mediterranean. *Dynamics of Atmospheres and Oceans*, 15:179–214, 1991.

30. J. C. J. Nihoul. Do not use a simple model when a complex one will do. *J. Mar. Syst.*, 5:401–406, 1994.

31. C. Parés, J. Macías, and M. J. Castro. Duality methods with an automatic choice of parameters. Application to shallow-water equations in conservative form. *Numer. Math.*, 89(1):161–189, 2001.

32. H. Perkins, T. H. Kinder, and P. E. La Violette. The Atlantic inflow in the western Alboran Sea. *J. Phys. Oceanogr.*, 20:242–263, 1990.

33. O. Pironneau. *Méthodes des Éléments Finis pour les Fluides*, volume 7 of *RMA*. Masson, 1988.

34. R. H. Preller. A numerical model study of the Alboran Sea gyre. *Progr. in Oceanogr.*, 16:113–146, 1986.

35. J. L. Reid. On the contribution of the Mediterranean Sea outflow to the Norwegian Greenland Sea. *Deep Sea Res.*, 26:1199–1223, 1977.

36. S. Speich. *Étude du Forçage de la Circulation Océanique par les Détroits: Cas de la Mer d'Alboran*. PhD thesis, Université Paris VI, November 1992. 245 pp.

37. P. Tchernia. Océanographie régionale, description physique des océans et des mers. Centre d'edition et de documentation, ENSTA, 1978. 277 pp.

38. J. Tintoré, D. Gomis, S. Alonso, and G. Parrilla. Mesoscale dynamics and vertical motion in the Alboran Sea. *J. Phys. Oceanogr.*, 21(6):811–823, June 1991.

39. J. Tintoré, P. E. La Violette, I. Blade, and A. Cruzado. A study of an intense density front in the eastern Alboran Sea. The Almeria-Oran front. *J. Phys. Oceanogr.*, 18(10):1384–1397, 1988.

40. A. Viúdez, J. Tintoré, and R. L. Haney. Circulation in the Alboran Sea as determined by quasi-synoptic hydrographic observations. Part I : three-dimentional structure of the two anticyclonic gyres. *J. Phys. Oceanogr.*, 26:684–705, 1996.

41. F. E. Werner, A. Cantos-Figuerola, and G. Parrilla. A sensitivity study of reduced-gravity channel flows with applications to the Alboran Sea. *J. Phys. Oceanogr.*, 18:373–383, 1988.

42. J. A. Whitehead and A. R. Miller. Laboratory simulation of the gyre in the Alboran Sea. *J. Geophys. Res.*, 84:3733–3742, 1979.

Simulation of reactive transport in groundwater.
A comparison of two calculation methods

Maarten W. Saaltink and Jesús Carrera

Dep. d'Enginyeria del Terreny i Cartogràfica, ETSECCPB, Universitat Politècnica de Catalunya, c/Jordi Girona 1-3, Mòdul D-2, 08034 Barcelona, Spain.
{dsaaltink,carrera}@etseccpb.upc.es

Abstract. Numerical simulation of reactive transport in groundwater (that is, transport of solutes undergoing chemical reactions) requires the solution of a large number of mathematical equations, which can be highly non linear. The choice of a method to solve these equations may effect significantly both computation time and numerical behavior of the solution. Two types of methods exist: The Direct Substitution Approach (DSA), based on Newton-Raphson, and the Picard or Sequential Iteration Approach (SIA). The advantage of the DSA is that it converges faster and is more robust than the SIA. Its disadvantage is that one has to solve simultaneously a much larger set of equations than for the SIA. We applied both methods to several examples and compared computational behavior. Results showed that, for chemically difficult, cases, the SIA may require very small time steps leading to excessive computation times. The DSA displays a much more robust behavior, with computation times much less sensitive to the value of chemical parameters and generally smaller than the SIA.

1 Introduction

The use of numerical models can greatly help the performance assessment of waste disposal facilities, the study of groundwater contamination and the understanding of groundwater quality in natural systems and the processes undergone by rocks. These models should consider the concentrations of several species and should be able to simulate both solute transport processes, such as advection and dispersion, and chemical reactions, such as complexation, adsorption and precipitation. This requires the solution of a large number of mathematical equations, which can be highly non linear. For complex problems this may easily lead to excessive computation times. Therefore, the choice of an approach to solve these mathematical equations is important. Several approaches are available. However, one can consider them to be variants of two.

The first one is the Picard method that includes the Sequential Iteration Approach (SIA) and the Sequential Non Iteration Approach (SNIA). It consists of separately solving the chemical equations and the transport equations. The difference between the SIA and the SNIA is that the first iterates between these two types of equations, whereas the second does not. The SIA has been used by, among others, *Kinzelbach*

[1991], *Yeh and Tripathi* [1991], *Engesgaard and Kipp* [1992], *Šimunek and Suarez* [1994], *Zysset et al.* [1994], *Morrison et al.* [1995], *Schäfer and Therrien* [1995] and *Stollenwerk* [1995]. The SNIA has been used by, among others, *Liu and Narasimhan* [1989a], *McNab and Narasimhan* [1994], *Walter et al.* [1994], *Engesgaard and Traberg* [1996]. *Valocchi and Malmstead* [1992], *Miller and Rabideau* [1993] and *Barry et al.* [1996] discussed some limitations of the SNIA and proposed solutions.

The second approach is the Newton-Raphson method, also called one-step, global implicit or Direct Substitution Approach (DSA). It consists of substituting the chemical equations into the transport equations and solving them simultaneously, applying Newton-Raphson. It has been used by, among others, *Valocchi et al.* [1981], *Steefel and Lasaga* [1994], *White* [1995], *Grindrod and Takase* [1996] and *Saaltink et al.* [1998].

The main disadvantage of the DSA is the large set of equations that one has to solve simultaneously, leading to high computational costs per iteration. We should also mention that programming the DSA is significantly more difficult than the SIA. On the other hand, the SIA and SNIA generally show slower convergence and are less robust and more stiff. This may require finer temporal discretisations than the DSA, leading to a larger number of iterations. *Reeves and Kirkner* [1988] and *Steefel and MacQuarrie* [1996] compared the different approaches by applying them to a number of cases of small one-dimensional grids. Both reported more numerical problems for the SIA and/or SNIA than for the DSA. The first found generally smaller computation times for the DSA, whereas the latter for the SIA and SNIA. Nevertheless, in both articles the computation times for the different approaches were always of the same order of magnitude. In an article which has had great impact, *Yeh and Tripathi* [1989] compared the different approaches for larger grids of one, two and three dimensions. They concluded that the DSA leads to excessive CPU memory and CPU times of realistic two- and three-dimensional cases, due to the very large set of equations that one has to solve for the DSA in these cases. However, they made their comparison on a theoretical basis without in fact applying them and measuring CPU times. Therefore, they could not take into account the fact that the DSA may require fewer iterations. We conjecture that for some cases this may be important and that hence the DSA might become more advisable than stated by *Yeh and Tripathi* [1989].

The objective of our work is precisely to test this conjecture. To do so, we first formulate.several cases of varying chemical complexity, second, solve them with the SIA and DSA and, third, compare required temporal discretisation, number of iteration and CPU time.

We start by explaining the mathematical formulation for reactive transport. The next section treats the implementation of SIA and DSA. Then, we give a short description of the cases that we used for the comparison. The next section discusses the results of the comparison between the SIA and DSA. Finally, the last section contains some conclusions.

2 Basic Equations

In this chapter we briefly explain the mathematical formulation for reactive transport. For a more detailed explanation we refer to *Saaltink et al.* [1998]. There are two types of equations that one has to solve: equations that express the chemical reactions and those that express mass balances and transport processes such as advection, dispersion and diffusion.

2.1 Chemical Reactions

If a system is in chemical equilibrium one can apply the mass action law that relates the concentration of reactant and products of a chemical reaction. This can be written for the whole system in the following form:

$$\mathbf{S_e}\left(\log \mathbf{c} + \log \boldsymbol{\gamma}(\mathbf{c})\right) = \log \mathbf{k} \tag{1}$$

where $\mathbf{S_e}$ is a $N_r \times N_s$ matrix (N_r being the number of reactions and N_s the number of chemical species) containing the stoichiometric constants of the reactions (i.e., the number of moles supplied/consumed in each reactions), $\mathbf{c}$ is a vector of the concentrations of all chemical species, $\mathbf{k}$ is a vector of equilibrium constants and $\boldsymbol{\gamma}$ a vector of activity coefficients which are a function of all concentrations. A special case are the minerals. One normally assumes that their activity (the product of the activity coefficient and concentration) always equals one.

The mass action law only applies at equilibrium. In other cases, slow chemical reactions are characterized by the reaction rate (r_k), which is defined as the amount of reactants evolving to products of a chemical reaction per unit time. It depends on the concentrations of species involved in the reaction but it may also depend on the concentration of catalysts, on the reactive surface (e.g., for precipitation/dissolution of minerals), on the amount of bacteria (for biological reactions), etc. In this work, we simply state that the reaction rate is a function of all concentrations:

$$\mathbf{r_k} = \mathbf{r_k}(\mathbf{c}) \tag{2}$$

2.2 Transport Equations

The basic equations for reactive transport can be written as follows:

$$\frac{\partial \mathbf{c}}{\partial t} = \mathbf{M}L(\mathbf{c}) + \mathbf{S_e^t r_e} + \mathbf{S_k^t r_k}(\mathbf{c}) \tag{3}$$

where $\mathbf{M}$ is a diagonal matrix that specifies whether a species is immobile or not, $\mathbf{S_k}$ is the stoichiometric matrix for kinetic reactions, $\mathbf{r_e}$ is the vector of reaction rates for equilibrium reactions and L is a linear operator for the convection, dispersion and prescribed sink/source terms. Notice that equation 3 simply expresses the contribution to the change in the concentration of all species ($\partial \mathbf{c}/\partial t$) caused by transport processes

(ML(c)), all equilibrium reactions ($S_e'r_e$) and all slow reactions ($S_k'r_k$). It is worth mentioning also that, while r_k can be written explicitly as a function of concentrations (equation 2), equilibrium reaction rates, r_e, cannot. They can only be expressed implicitly by coupling equations 1 and 3. For the sake of simplicity, we will consider only transport in a single, aqueous phase. Then, $L(c)$ is given by:

$$L(c) = -\frac{1}{\phi} \nabla \cdot (\mathbf{q}c) + \nabla \cdot (\mathbf{D}\nabla c) + m \tag{4}.$$

where $\mathbf{q}$ is the water flux, ϕ the volumetric water content, $\mathbf{D}$ the dispersion tensor and m represents sources and sinks.

As there are N_s concentrations per node, there are also N_s transport equations per node. We can reduce the number of transport equations by eliminating r_e. In order to do so, one should recall that S_e is a full-ranked $N_r \times N_s$ matrix. Full rank is assured because all equilibrium reactions must be independent. Therefore, it is possible to obtain a full ranked $(N_s - N_r) \times N_s$ kernel matrix ($\mathbf{U}$) such that:

$$\mathbf{U}S_e' = \mathbf{0} \Rightarrow \mathbf{U}S_e'r_e = \mathbf{0} \tag{5}$$

Multiplying equation 4 by $\mathbf{U}$ allows us to eliminate the equilibrium reaction rates term r_e:

$$\mathbf{U}\frac{\partial \mathbf{c}}{\partial t} = \mathbf{U}ML(\mathbf{c}) + \mathbf{U}S_k'r_k(\mathbf{c}) \tag{6}$$

Since the dimensions of $\mathbf{U}$ are $(N_s - N_r) \times N_s$, the number of transport equations per node reduces from N_s in equation 4 to $N_s - N_r$ in equation 6. We will call $\mathbf{U}$ the component matrix, because it adds up the total amount of a component, distributed over the various chemical species. Components are defined in such a way, that every species can be uniquely represented as a combination of one or more components [*Yeh and Tripathi*,1989]. In addition, equation 5 ensures that the global mass of a component is independent of equilibrium reactions [*Rubin*, 1983]. In a closed system the global mass only depends on kinetic reactions, whereas in an open system the global mass depends on mass fluxes as well.

Due to the assumption that the activity of minerals equals one, we can also eliminate the concentrations of these species. To do this, we write equation 6 in the following form:

$$\mathbf{U}_a \frac{\partial \mathbf{c}_a}{\partial t} + \mathbf{U}_s \frac{\partial \mathbf{c}_s}{\partial t} + \mathbf{U}_m \frac{\partial \mathbf{c}_m}{\partial t} = \mathbf{U}_a L(\mathbf{c}_a) + \mathbf{U}S_k'r_k(\mathbf{c}_a) \tag{7}$$

where vector $\mathbf{c}$ and matrices $\mathbf{U}$ have been split up into parts referring to the aqueous and therefore mobile species (with subscript $\mathbf{a}$), sorbed and therefore immobile species (with subscript $\mathbf{s}$) and minerals (with subscript $\mathbf{m}$):

$$\mathbf{c} = \begin{pmatrix} \mathbf{c_a} \\ \mathbf{c_s} \\ \mathbf{c_m} \end{pmatrix} \tag{8}$$

$$\mathbf{U} = \begin{pmatrix} \mathbf{U_a} & \mathbf{U_s} & \mathbf{U_m} \end{pmatrix} \tag{9}$$

In the same way as for the elimination of equilibrium reaction rates, we eliminate $\mathbf{U_m}\partial\mathbf{c_m}/\partial t$ by multiplying equation 7 by an elimination matrix $\mathbf{E}$ defined in such a way that:

$$\mathbf{EU_m} = \mathbf{0} \Rightarrow \mathbf{EU_m}\frac{\partial\mathbf{c_m}}{\partial t} = \mathbf{0} \tag{10}$$

Then, equation 7 becomes:

$$\mathbf{EU_a}\frac{\partial\mathbf{c_a}}{\partial t} + \mathbf{EU_s}\frac{\partial\mathbf{c_s}}{\partial t} = \mathbf{EU_a}L(\mathbf{c_a}) + \mathbf{EUS}_k^t\mathbf{r_k}(\mathbf{c_a}) \tag{11}$$

Multiplying by $\mathbf{E}$ reduces the number of transport equations from $N_s - N_r$ in equation 7 to $N_s - N_r - N_m$ in equation 11 (N_m being the number of minerals in equilibrium).

3 Numerical Approaches

3.1 Sequential Iteration Approach (SIA)

We use equation 7 but written in a slightly different form:

$$\frac{\partial\mathbf{u_a}}{\partial t} + \frac{\partial\mathbf{u_s}}{\partial t} + \frac{\partial\mathbf{u_m}}{\partial t} = L(\mathbf{u_a}) + \mathbf{US}_k^t\mathbf{r_k}(\mathbf{u_a}) \tag{12}$$

where $\mathbf{u_a}$, $\mathbf{u_s}$ and $\mathbf{u_m}$ are vectors containing the total concentrations of the component in aqueous, sorbed and mineral form respectively. They are defined as:

$$\mathbf{u_a} = \mathbf{U_a}\mathbf{c_a} \tag{13}$$

$$\mathbf{u_s} = \mathbf{U_s}\mathbf{c_s} \tag{14}$$

$$\mathbf{u_m} = \mathbf{U_m}\mathbf{c_m} \tag{15}$$

The SIA consists of first solving the transport equations 12 with the total aqueous concentrations of every chemical component (vector $\mathbf{u_a}$) as unknowns. It treats the concentrations of sorbed species, minerals and kinetic reactions as source-sink term (vector $\mathbf{f}$), computed by the previous iteration:

$$\frac{\partial \mathbf{u}_a^i}{\partial t} = L(\mathbf{u}_a^i) + \mathbf{f}^{i-1} \tag{16}$$

where the superscript i refers to the iteration number. Note that 16 has the same form as the transport equation without chemical reactions. We used finite elements for spatial and finite differences for the temporal discretization. This leads to linear equations that one can solve for every component separately. We used LU decomposition of a banded matrix to solve these linear equations. As the matrix of this system only changes if the time increment changes, one can take advantage of the decomposed matrix to solve the systems of all components and even of previous time steps as long as the time increment is not changed.

In the second step one updates the source-sink terms. As there is no explicit expression for $\mathbf{f}$ as a function of $\mathbf{u}_a$, one has to calculate first the concentrations ($\mathbf{c}$) from the total aqueous concentrations ($\mathbf{u}_a$) by means of the chemical equations:

$$\mathbf{u}_a^i + \mathbf{u}_d^{i-1} + \mathbf{u}_m^{i-1} = \mathbf{U}_a \mathbf{c}_a + \mathbf{U}_d \mathbf{c}_d + \mathbf{U}_m \mathbf{c}_m \tag{17}$$

One has to solve these equations together with equations 14 to 16 and those for chemical equilibrium (equation 1). One usually substitutes the chemical equations into 17 with the exception of the equations for reactions that involve minerals. This leads to $N_s - N_r + N_m$ number of equations. Because these equations are non linear, we applied a Newton-Raphson scheme for its solution (not to be confused with the Newton-Raphson applied to the DSA to solve the whole set of equations). One can do this for every node separately. From $\mathbf{c}$ one calculates new source-sink terms:

$$\mathbf{f}^i = \mathbf{U}_d \frac{\partial \mathbf{c}_d}{\partial t} + \mathbf{U}_m \frac{\partial \mathbf{c}_m}{\partial t} + \mathbf{U} \mathbf{S}_k^t \mathbf{r}_k(\mathbf{c}) \tag{18}$$

This term can now be substituted in (16) for the next iteration and the whole process is repeated until convergence.

3.2 Direct Substitution Approach (DSA)

For the DSA we substitute all chemical equations (1) into the transport equation (11) and apply Newton-Raphson as follows. We define a vector of transport equations ($\mathbf{g}$) and of chemical equations ($\mathbf{h}$):

$$\mathbf{g} = \mathbf{EU}_a \frac{\partial \mathbf{c}_a}{\partial t} + \mathbf{EU}_s \frac{\partial \mathbf{c}_s}{\partial t} - \mathbf{EU}_a L(\mathbf{c}_a) - \mathbf{EU} \mathbf{S}_k^t \mathbf{r}_k(\mathbf{c}_a) = 0 \tag{19}$$

$$\mathbf{h} = \mathbf{S}_e \big(\log \mathbf{c} + \log \gamma(\mathbf{c})\big) - \log \mathbf{k} = 0 \tag{20}$$

We also define a vector of $N_s - N_r - N_m$ concentrations (c_1) which we will link to the transport equations and another of N_r concentrations (c_2) linked to the chemical equations. Then the Newton-Raphson scheme becomes:

$$\frac{\partial \mathbf{g}}{\partial \mathbf{c}_1}\left(\mathbf{c}_1^{i+1} - \mathbf{c}_1^i\right) + \frac{\partial \mathbf{g}}{\partial \mathbf{c}_2}\left(\mathbf{c}_2^{i+1} - \mathbf{c}_2^i\right) = -\mathbf{g}^i \tag{21}$$

$$\frac{\partial \mathbf{h}}{\partial \mathbf{c}_1}\left(\mathbf{c}_1^{i+1} - \mathbf{c}_1^i\right) + \frac{\partial \mathbf{h}}{\partial \mathbf{c}_2}\left(\mathbf{c}_2^{i+1} - \mathbf{c}_2^i\right) = -\mathbf{h}^i \tag{22}$$

where $\mathbf{h}^i$ and $\mathbf{g}^i$ stand for $\mathbf{h}(\mathbf{c}^i)$ and $\mathbf{g}(\mathbf{c}^i)$, respectively. If we ensure concentrations c_2 to be in equilibrium with c_1, chemical equations ($\mathbf{h}$) equal zero. To fulfill this condition we applied the following Newton-Raphson scheme to calculate concentrations c_2 for given concentrations c_1:

$$\frac{\partial \mathbf{h}}{\partial \mathbf{c}_2}\left(\mathbf{c}_2^{j+1} - \mathbf{c}_2^j\right) = -\mathbf{h}^j \tag{23}$$

We substitute 22 into 21 with chemical equations ($\mathbf{h}$) being zero:

$$\left(\frac{\partial \mathbf{g}}{\partial \mathbf{c}_1} + \frac{\partial \mathbf{g}}{\partial \mathbf{c}_2}\frac{\partial \mathbf{c}_2}{\partial \mathbf{c}_1}\right)\left(\mathbf{c}_1^{i+1} - \mathbf{c}_1^i\right) = -\mathbf{g}^i \tag{24}$$

where $\partial \mathbf{c}_2 / \partial \mathbf{c}_1$ is defined by:

$$\frac{\partial \mathbf{h}}{\partial \mathbf{c}_2}\frac{\partial \mathbf{c}_2}{\partial \mathbf{c}_1} = -\frac{\partial \mathbf{h}}{\partial \mathbf{c}_1} \tag{25}$$

The approach consist of first calculating c_2 by means of equation 23. Then we calculate $\partial \mathbf{c}_2 / \partial \mathbf{c}_1$ from 25. Note, that equation 23 and 25 have the same jacobian matrix ($\partial \mathbf{h} / \partial \mathbf{c}_2$) and that they are local, that is, they represent equilibrium conditions at every point, so that they can be solved for every node separately. Moreover, as for normal situations in groundwater γ generally varies only slightly with concentrations, we can write the chemical equations ($\mathbf{h}$) almost as an explicit function of c_1 ($c_2 = \mathbf{f}(c_1)$) [Saaltink et al., 1998], which leads a jacobian matrix equal to the unity matrix. Therefore, calculation of 23 and 25 is not extraordinarily costly. After calculation of c_2 and $\partial \mathbf{c}_1 / \partial \mathbf{c}_2$ we calculate a new value of c_1 by means of equation 24. This equation has $N_s - N_r - N_m$ unknowns that must be solved for all nodes simultaneously.

As for the SIA, we used finite elements for spatial and finite difference for the temporal discretization.

3.3 Time increment control

Both the SIA and DSA can fail to converge if one uses a too large time increment. On the other hand, CPU times can become unnecessarily large if one uses a too small time increment. As we conjecture, that the SIA requires smaller time increments than the DSA, it is important to work with the optimal time increments for both approaches. However, it is difficult to estimate a priori an optimal time increment. Therefore, we developed an algorithm that changes automatically the time increment during the simulation, according to the following scheme:

```
IF (failed to converge) THEN
  Decrease time step DT by a factor FD
ELSE
  IF (number of iter. < min. threshold THRMIN) THEN
    Increase DT by a factor FI
    IF (DT > maximum time increment DTMAX) THEN
      DT = DTMAX
    ENDIF
  ELSEIF (number of iter. > max. threshold THRMAX) THEN
    decrease DT by factor FD
  ELSE
    maintain DT
  ENDIF
ENDIF
Do next time step
```

If convergence cannot not be reached, it repeats the calculations with a smaller time step. In case of a successful convergence, it reduces the time increment for the next time step, if the number of iterations is large, whereas the time increment is increased, if this number is small.

4 Case descriptions

Table 1 shows a summary of the cases used for the comparison. The first set of examples (CAL) is the most simple one. It treats the dissolution of calcite in a one dimensional domain. Initially the water is saturated with calcite. Infiltrating water, that is subsaturated to calcite, dissolves the calcite. This case consists of several subcases: one assuming equilibrium dissolution of calcite (CAL-E) and four with kinetic dissolution, with various calcite dissolution rates (CAL-1 to CAL-4, the first having the slowest rate and the last the fastest). We also calculated a case without calcite (CAL-0).

The next set (WAD) contains cases of the flushing of saline water by fresh water in the Waddenzee (the Netherlands) in a one dimensional domain described by *Appelo and Postma* [1994]. They include dissociation of water, carbonate reactions, cation exchange and dissolution of calcite. Likewise the calcite dissolution cases, we calculated cases of equilibrium, kinetic and no calcite dissolution.

The third set (DEDO) simulates the replacement of dolomite with calcite, which is driven by the infiltration of Ca rich water, called dedolomitization [*Ayora et al.*, 1998]. We used a two dimensional domain, which includes a fracture with a 100 times higher water velocity than in the surrounding rock. Note that a high number of pore volumes are flushed for this case. The number of flushed pore volumes is the volume of water that enters the domain during the simulated time divided by the volume of water in the domain.

The last case (OSA) is chemically the most complex one. It models the deposition of uranium resulting from infiltration of oxygenated, uranium bearing groundwater through a hydrothermally altered phonolitical host rock at the Osamu Utsumi uranium mine, Poços de Caldas, Brazil [*Lichtner and Waber*, 1992]. As for the DEDO cases, a high number of pore volumes are flushed.

Table 1. Characteristics of the cases used for comparison.

Case name	No. of nodes	No. of primary species	No. of second. species	No. of adorbed species	No. of minerals (equil.)	No. of minerals (kin.)	Flushed pore volumes
CAL-0	21	3	5	-	-	-	1.0
CAL-1	21	3	5	-	-	1	1.0
CAL-2	21	3	5	-	-	1	1.0
CAL-3	21	3	5	-	-	1	1.0
CAL-4	21	3	5	-	-	1	1.0
CAL-E	21	3	5	-	1	-	1.0
WAD-0	21	6	3	3	-	-	37.5
WAD-1	21	6	3	3	-	1	37.5
WAD-2	21	6	3	3	-	1	37.5
WAD-3	21	6	3	3	-	1	37.5
WAD-4	21	6	3	3	-	1	37.5
WAD-E	21	6	3	3	1	-	37.5
DEDO-E	15×15	7	8	-	2	-	22704.6
DEDO-K	15×15	7	8	-	-	2	22704.6
OSA	101	13	29	-	8	-	80000.0

5 Comparison

We measured the number of required time steps, the number of required iterations and the CPU times are shown in figures 1 through 3, respectively, for the cases described above. The numbers of time steps and iterations may depend on the parameters that control the time increment and convergence criteria and the CPU time on the programming style. So one should interpret these figures with care. Nevertheless, one can observe some clear differences between the two approaches. For the cases with a small number of flushed pore volumes (CAL and WAD), the SIA and the DSA behave similarly, when the mineral is in equilibrium or absent. The number of time steps and iterations and consequently the CPU time of the SIA rise with higher dissolution rates, when one assumes a kinetic dissolution. On the other hand, dissolution rates do not seem to have much influence on the numerical behaviour of the DSA. A bigger kinetic rate makes the non-linear source-sink term **f** in equation 10 to be more important, causing more numerical problems for the SIA. The DSA does not show these problems thanks to its robustness.

For the cases with a high number of flushed pore volumes (DEDO, OSA) the SIA requires really excessive number of time steps, iteration and CPU time (centuries) also for cases that assume equilibrium dissolution-precipitation (in fact, we have estimated these figures by extrapolating from runs that took about one day). The high number of flushed pore volumes makes that fulfilling the Courant condition would lead to a very high number of time step. This condition states that the solute cannot go over an element during a simple time step:

$$\Delta t \leq \frac{\phi \Delta x}{q} \tag{26}$$

where Δx and Δt are the element size and the time increment. It seems that the Courant condition is important for the SIA, whereas it is not for the DSA.

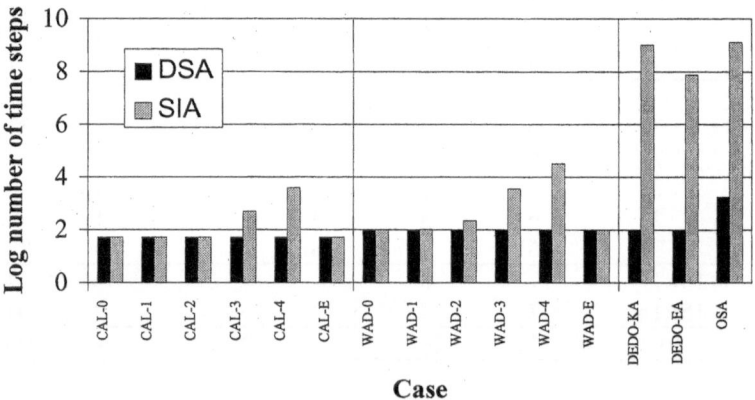

Fig. 1. Number of time steps for all cases. The values for the DEDO and OSA cases calculated by the SIA have been derived by extrapolating those for a small number of time steps. Note the log scale.

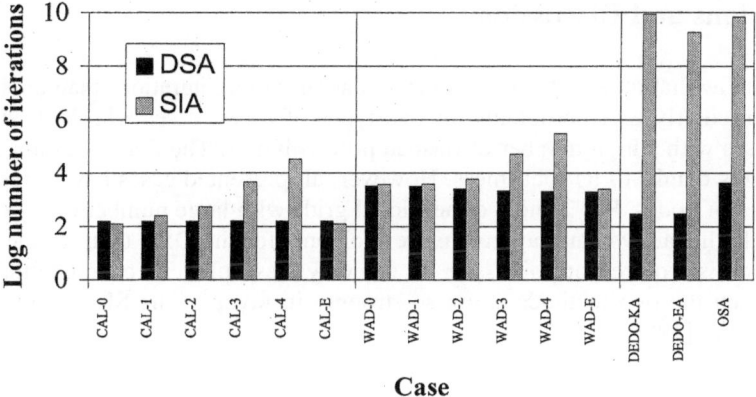

Fig. 2. Number of iterations for all cases. The values for the DEDO and OSA cases calculated by the SIA have been derived by extrapolating those for a small number of time steps. Note the log scale.

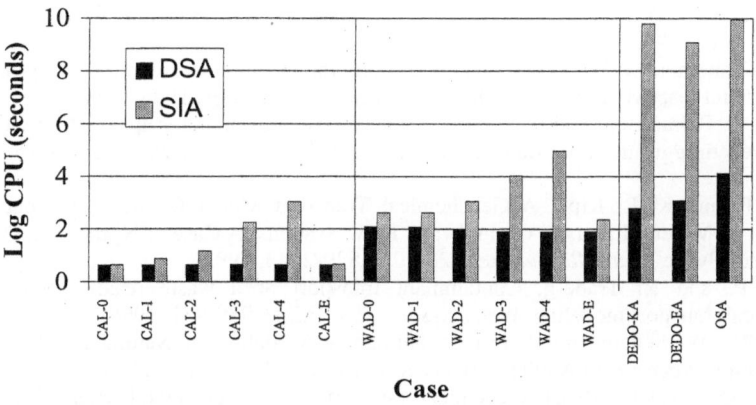

Fig. 3. CPU times for all cases. The values for the DEDO and OSA cases calculated by the SIA have been derived by extrapolating those for a small number of time steps. Note the log scale.

6 Conclusions and Discussion

The results show that indeed the SIA requires generally more iterations than the DSA. The SIA particularly gives problems for two types of cases: cases with high kinetic rates and cases with a high number of flushed pore volumes. The DSA does not show these problems thanks to its robustness. However, all presented cases have grids of a small number of nodes. For 2 and 3 dimensional grids with large number of nodes the solution of the linear system may give more problems for the DSA than for the SIA. More research is required on this issue. At the very least, one can conclude that the choice between the two methods is not so strongly in favor of the SIA as stated by *Yeh and Tripathi* [1989].

Acknowledgments

This work was funded by ENRESA (Spanish Nuclear Waste Management Company).

References

Appelo, C. A. J. and D. Postma, *Geochemistry, Groundwater and Pollution*, A. A. Balkema, Brookfield, Vt., 1994.

Ayora, C., C. Taberner, M. W. Saaltink and J. Carrera, the origin of dedolomites. A discussion on textures and reactive transport modeling, *Journal of Hydrology*, 1998, in press.

Barry, D. A., K. Bajracharya and C. T. Miller, Alternative split-operator approach for solving chemical reaction/groundwater transport models, *Advances in Water Resources, 19(5)*, 261-275, 1996.

Engesgaard, P. and K. L. Kipp, A Geochemical Transport Model for Redoxed-Controlled Movement of Mineral Fronts in Groundwater Flow Systems: A Case of Nitrate Removal by Oxidation of Pyrite, *Water Resour. Res., 28(10)*, 2829-2843, 1992.

Engesgaard, P. and R. Traberg, Contaminant transport at a waste residue deposit. 2. Geochemical transport modeling, *Water Resour. Res., 32(4)*, 939-951, 1996.

Kinzelbach, W., W. Schäfer and J. Herzer, Numerical Modeling of Natural and Enhanced Denitrification Processes in Aquifers, *Water Resour. Res., 27(6)*, 1123-1135, 1991.

Grindrod, P. and H, Takase, Reactive chemical transport within engineered barriers, *Journal of Contaminant Hydrology, 21*, 283-296, 1996.

Liu, C. W. and T. N. Narasimhan, Redox-Controlled Multiple-Species Reactive Chemical Transport, 1. Model Development, *Water Resour. Res., 25(5)*, 869-882, 1989a.

Liu, C. W. and T. N. Narasimhan, Redox-Controlled Multiple-Species Reactive Chemical Transport, 2. Verification and Application, *Water Resour. Res., 25(5)*, 883-910, 1989b.

Lichtner, C. L. and N. Waber, Redox front geochemistry and weathering: theory with application to the Osamu Utsumi uranium min, Poços de Caldas, Brazil, *Journal of Geochemical exploration, 45*, 521-564, 1992.

McNab Jr., W.W. and T. N. Narasimham, Modeling reactive transport of organic compounds in groundwater using a partial redox disequilibrium approach, *Water Resour. Res., 30(9)*, 2619-2635, 1994.

Miller, C. T. and A. J. Rabideau, Development of Split-Operator, Petrov-Galerkin Methods to Simulate Transport and Diffusion Problems, *Water Resour. Res., 29(7)*, 2227-2240, 1993.

Morrison, S. T., V. S. Tripathi and R. R. Spangler, Coupled reaction/transport modeling of a chemical barrier for controlling uranium(VI) contamination in groundwater, *Journal of Contaminant Hydrology, 17*, 347-363, 1995.

Rubin, J., Transport of Reacting Solutes in Porous Media: Relation Between Mathematical Nature of Problem Formulation and Chemical Nature of Reactions, *Water Resour. Res., 19(5)*, 1231-1252, 1983.

Saaltink, M. W., C. Ayora and J. Carrera, A mathematical formulation for reactive transport that eliminates mineral concentrations, *Water Resour. Res., 34(7)*, p.1649-1656, 1998.

Schäfer, W. and R. Therrien, Simulating transport and removal of xylene during remediation of a sandy aquifer, *Journal of Contaminant Hydrology, 19*, 205-236, 1995.

Šimunek, J. and D. L. Suarez, Two-dimensional transport model for variability saturated porous media with major ion chemistry, *Water Resour. Res., 30(4)*, 1115-1133, 1994.

Steefel, C. I. and A. C. Lasaga, A Coupled Model for Transport of Multiple Chemical Species and Kinetic Precipitation/Dissolution Reactions with Application to Reactive Flow in Single Phase Hydrothermal Systems, *American Journal of Science, 294*, 529-592, 1994.

Stollenwerk, K. G., Modeling the effects of variable groundwater chemistry on adsorption of molybdate, *Water Resour. Res., 31(2)*, 347-357, 1995.

Valocchi, A. J., R. L. Street and P. V. Roberts, Transport of Ion-Exchanging Solutes in Groundwater: Chromatographic Theory and Field Simulation, *Water Resour. Res., 17(5)*, 1517-1527, 1981.

Valocchi, A. J. and M. Malmstead, Accuracy of Operator Splitting for Advection-Dispersion-Reaction Problems, *Water Resour. Res., 28(5)*, 1471-1476, 1992.

Walter, A. L., E. O. Frind, D. W. Blowes, C. J. Ptacek and J. W. Molson, Modeling of multicomponent reactive transport in groundwater, 1. Model development and evaluation, *Water Resour. Res., 30(11)*, 3137-3148, 1994.

White, S. P., Multiphase nonisothermal transport of systems of reacting chemicals, *Water Resour. Res., 31(7)*, 1761-1772, 1995.

Yeh, G. T. and V. S. Tripathi, A Critical Evaluation of Recent Developments in Hydrogeochemical Transport Models of Reactive Multichemical Components, *Water Resour. Res., 25(1)*, 93-108, 1989.

Yeh, G. T. and V. S. Tripathi, A Model for Simulating Transport of Reactive Multispecies Components: Model Development and Demonstration, 27(12), 3075-3094, *Water Resour. Res., 27(12)*, 1991.

Zysset, A., F. Stauffer and T. Dracos, Modeling of chemically reactive groundwater transport, *Water Resour. Res., 30(7)*, 2217-2284, 1994.